BIOMEDICAL IMAGE ANALYSIS: TRACKING

Biomedical Image Analysis: Tracking

Scott T. Acton and Nilanjan Ray

ISBN: 978-3-031-01109-2   Acton/Ray, Biomedical Image Analysis: Tracking (paperback)
ISBN: 978-3-031-02237-1   Acton/Ray, Biomedical Image Analysis: Tracking (e-book)
DOI 10.1007/978-3-031-02237-1

Library of Congress Cataloging-in-Publication Data

First Edition
10 9 8 7 6 5 4 3 2 1

BIOMEDICAL IMAGE ANALYSIS: TRACKING

Scott T. Acton and Nilanjan Ray
University of Virginia

Alan C. Bovik, *Series Editor*

ABSTRACT

In biological and medical imaging applications, tracking objects in motion is a critical task. This book describes the state-of-the-art in biomedical tracking techniques. We begin by detailing methods for tracking using active contours, which have been highly successful in biomedical applications. The book next covers the major probabilistic methods for tracking. Starting with the basic Bayesian model, we describe the Kalman filter and conventional tracking methods that use centroid and correlation measurements for target detection. Innovations such as the extended Kalman filter and the interacting multiple model open the door to capturing complex biological objects in motion.

A salient highlight of the book is the introduction of the recently emerged particle filter, which promises to solve tracking problems that were previously intractable by conventional means. Another unique feature of *Biomedical Image Analysis: Tracking* is the explanation of shape-based methods for biomedical image analysis. Methods for both rigid and nonrigid objects are depicted. Each chapter in the book puts forth biomedical case studies that illustrate the methods in action.

KEYWORDS

Active contours, Biomedical imaging, Image analysis, Image processing, Medical imaging, Particle filter, Snakes, Kalman filter, Target tracking.

To Gracene Acton and Rupa Ray

Contents

C H A P T E R 1

Introduction

"The world leaves no track in space, and the greatest action of man no mark in the vast idea."

—Henry David Thoreau

"Never mistake motion for action."

—Ernest Hemingway

Excerpts from an interview by Hadley Marie with the authors conducted in May 2005:

Q: *What makes this book different from other books on tracking?*

A: Other books on tracking focus on the controls and estimation theory aspects of tracking. This book puts forth methods for extracting image information from biological/medical images and for using this information in tracking biological targets.

Q: *How does the book differ from the host of medical imaging books currently available?*

A: Here, and in the forthcoming companion book *Biomedical Image Analysis: Segmentation* (Morgan-Claypool Pub.), we concentrate on aspects of image analysis rather than the modalities or the imaging process itself.

Q: *What makes biomedical image analysis and tracking unique?*

A: The reason for treating biomedical image analysis separately is that we are imaging and analyzing living organisms, which imply moving backgrounds and complex interfaces. These organisms, organs and tissue are deformable objects that can change contrast, that have subtle, ambiguous boundaries, that can be occluded, and that are imaged in the presence of significant clutter, noise and speckle.

In military imaging we have rigid bodies and distinct boundaries. In a manufacturing environment we are presented with high contrast rigid objects with distinct boundaries, controlled illumination, and controlled angles of presentation. The biomedical world is far more unpredictable and more difficult to constrain.

Finally, we must have knowledge of what is doing the imaging, in other words, the modalities. These modalities include ultrasound, microscopy, CT, MRI, PET/SPECT, X-Ray, EIT, and OPT.

Q: *What if I'm interested in tracking but not biology or medicine?*

A: This book could still be a valuable resource. The title *Biomedical Image Analysis: Tracking* reflects our favorite application area—an application area that we believe contains unique challenges in tracking.

Q: *Who is the intended audience for the book?*

A: Graduate students, faculty, and industrial/governmental researchers interested in applications of imaging, or more specifically, biomedical imaging. The book is written generally from first principles and should be accessible to a broad readership.

Q: *Is the book accessible by undergraduates?*

A: Yes, the precocious junior or average senior in biomedical engineering, electrical engineering, computer engineering, computer science, systems engineering, mechanical/aerospace engineering, civil engineering, physics, or mathematics can master this text. Essentially, we assume knowledge of undergraduate probability, basic linear systems, and calculus.

Q: *Is the information in the book solely about your research?*

A: No. Although we are biased in presenting the work that we know best (our work), we are excited to summarize other methods developed by many of today's most impacting experts in biomedical image analysis.

Q: *Does the book provide a balanced slate of biomedical applications that encompasses the entire spectrum of biomedical imaging?*

A: In a nutshell, no. The examples are processing and analysis techniques given in this book emerge mainly from cell tracking applications in microscopy. We do provide other examples of biomedical tracking including cardiac motion, vessel wall motion, and cartilage boundary tracking. However, the number of potential biomedical tracking applications precludes complete coverage from the application side. We are confident that the solutions limned in this book can be adapted to and adopted by a large number of these applications in biology and medicine.

Q: *What exactly is in the book?*

A: Chapter 2 develops and details methods for tracking using active contours, which have been highly successful in biomedical applications. Here, we treat the basic model and extended models with specialized constraints. We pay particular attention to the external force of the snake—the force that propels the contour toward the desired border. One of the main vulnerabilities of the active contour, the selection of parameters and weights, is discussed. Finally, methods are given for implementing snakes by way of dynamic programming, in addition to the traditional gradient descent implementation.

Chapter 3 of the book explores probabilistic methods for tracking. Starting with the basic Bayesian model, we describe the Kalman filter and conventional tracking methods that use centroid and correlation measurements for target detection. A comprehensible implementation of the alpha–beta filter is highlighted, and the extended Kalman filter is described. The interacting multiple model opens the door to simultaneous use of different motion models for the same biological object. The chapter concludes with a demonstration of the Kalman filter's application to biomedicine.

In Chapter 4 we explore factored sampling and Monte Carlo methods including the newly emerging particle filter. This chapter summarizes important new advances in the target-tracking arena, including multi-target tracking techniques. Biomedical examples are used to demonstrate the implementation of the particle filter and associated methods.

Chapter 5 looks at shape-based methods for biomedical image analysis. The first topic details an active contour approach based on iterative updates of affine and projective transformation parameters. For nonrigid shapes, we apply stochastic models, such as snakes driven by annealing methods. Finally, a Markov Chain Monte Carlo approach is outlined and the sequential Bayesian algorithm is revisited where the multi-target model is adapted to a shape-based approach.

Q: *Are there any acknowledgments you would like to make?*

A: Indeed. The authors would like to thank series editor Al Bovik and publisher Joel Claypool for their support and hard work connected with realizing this book. The authors are indebted to Klaus Ley and his laboratory at the University of Virginia for the intravital imagery used in this book. Other collaborators that have contributed imagery include John Hossack and Fred Epstein of the University of Virginia. Also, many members of Virginia Image and Video Analysis (VIVA) contributed to the techniques and demonstrations used in the book. The book was improved due to the comments of the reviewers, particularly those by Dr. Dipti Mukherjee, Dr. Peter Tay, Dr. Thomas Wuerflinger and editor Al Bovik. The authors also thank the National Institutes of Health for sponsoring their biomedical tracking research (HL68510, EB001763, EB001826).

CHAPTER 2

Active Contours for Tracking

"The earth doth like a snake renew Her winter weeds outworn."

—Percy Bysshe Shelley

"Which of you fathers, if your son asks for a fish, will give him a snake instead?"

—Jesus Christ

2.1 OVERVIEW

The flexible outline provided by an active contour (a.k.a. the snake) is mated well with the amorphous, nonrigid boundaries found in many biomedical applications. The purpose of this chapter is to put forth the active contour model as a tool for object tracking. Essentially, the active contour is a contour that seeks to optimize *energy*, where the energy quantifies the "goodness" of the contour—how smooth it is and how well localized it is with respect to the image edges. The movement of the active contour is then motivated by minimizing this energy, making moves that improve the goodness of the contour in some manner. We first discuss the concept of gradient descent as an energy minimization tool for active contour computation. Employment of the active contour for cell tracking is addressed next. In this context, we illustrate several external forces for active contours—the external forces that guide the active contour to reside on the desired image edges. Here, the *minimax* method for determining active contour parameter values is outlined. Finally, dynamic programming for snake energy minimization is discussed as an alternative to the gradient descent method.

2.2   THE BASIC SNAKE MODEL

In the eighties, disco died and the snake was born. And as disco band "The Bee Gees" had three founders, so the snake had its three creators: Kass, Witkin, and Terzopoulos [1]. After nearly three decades of deformation, biomedical imaging researchers have a mature tool that accommodates segmentation and tracking of nonrigid objects.

The basic snake model introduced by Kass *et al*. is a good starting point for exploration of the world of active contours. We are first going to introduce this basic snake model and then we will utilize the model to track objects of interest in biomedical images. The crux of the snake-based tracking will be using an active contour to lock onto the boundary of a moving object whilst the object moves in a video sequence.

A snake is simply the mathematical description of a contour used to delineate a boundary. As mentioned, the goal is moving the contour such that an energy measure is minimized. To understand the working principles of the snake model, let us first consider an image of a circle as shown in Fig. 2.1(a). Figure 2.1(b) shows the image intensity profile as a surface plot. The negative (additive inverse) of the image gradient magnitude is shown in Fig. 2.1(c), and corresponding surface plot is shown in Fig. 2.1(d). The additive inverse of the gradient magnitude forms a topographic surface and serves as the potential energy for the snake. Object boundaries correspond to the valleys of this topographic surface, *e.g.* the annular valley shown in Fig. 2.1(d). Let us consider point A on the circle image as shown in Fig. 2.1(e), which is a magnified version of Fig. 2.1(d). The potential energy at A may be considered as the height of the potential surface (negative gradient magnitude) at A, measured from some reference. According to the principle of

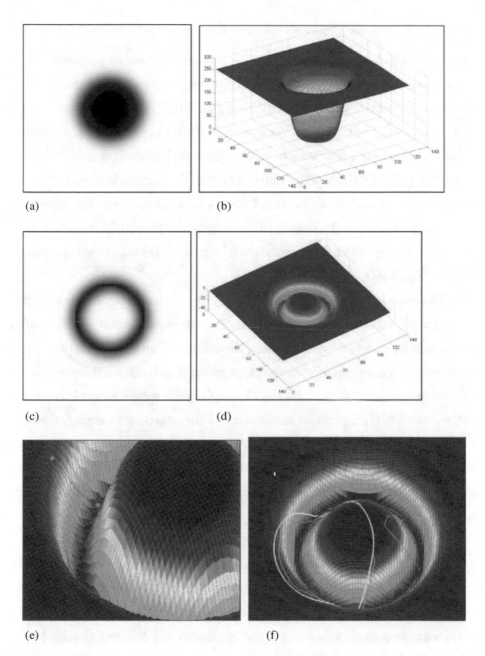

FIGURE 2.1: Illustration of the gradient descent method. (a) A circle image. (b) The circle image intensity as a surface. (c) Negative of the image gradient magnitude of the circle image. (d) Negative image gradient magnitude as a surface plot. (e) Points A and B on the negative gradient magnitude surface. The point A scrolls down to B along the path shown here. (f) Contours (collection of ordered points) on the negative gradient magnitude surface settle down to the valley

energy minimization, point A will always attempt to lower its potential energy; consequently, this point will amble down the potential surface and rest in a local minimum of the surface indicated by B (see Fig. 2.1(e)). A similar situation arises in the computation of the snake. We may approximate a contour (*viz.*, a snake) by a collection of points called *snaxels*. When a contour is placed on the circle image shown in Fig. 2.1(f), the total potential energy of the contour is given by the sum of potential surface heights (measured from a base reference) at the constituent snaxels. In this case too the snaxels migrate down the potential surface and reach the nearest local minima. This evolution of a contour on the potential surface is shown in Fig. 2.1(f).

The curious reader would immediately raise concerns regarding the collective behavior of the snaxels, for example whether the snaxels maintain smoothness of the contour while evolving or after reaching the minima. The good news preached in this chapter is that the inventors of the snake model already answer the question. We can reframe the question as, *is the movement of a snaxel influenced by the movements of other snaxels?* If independent movements of the snaxels are allowed, then it is very likely that after they reach their respective local minima, the contour shape would be jagged and sensitive to minor variations in the image gradient. In fact in biomedical imaging, this would almost always be the case since the potential surface is generally noisy.

So, the movement of a snaxel is typically dependent on the movements of a few neighboring snaxels so that the contour does not become rough or irregular. *How can one make this happen in the energy minimization framework?* The trick is to associate another kind of energy with the snake. This second energy function should assume a small value where the contour is smooth and should explode as the snake becomes jagged. Notice that now we have two energies associated with the snake—one is the potential energy computed from image data and the other is the snake smoothness energy. And we want both these energies to be minimized ideally. *But is it possible to achieve this goal of simultaneous minimization of the energies?* Certainly it is not guaranteed that a snake minimizes both the image potential energy and the smoothness energy—these energies may lead to conflicting constraints.

Thus, some sort of compromise has to be met between these two types of energies. In order to formally achieve this joint optimality, the total energy of the snake is considered to be the sum of the image potential energy and the smoothness energy. And we now want the snake to minimize this total energy. Again the curious and the cautious reader would ask whether the total energy minimization would lead to our desired contour location. In fact, we ask a more subtle question—*what is our "desired" snake location?* Intuitively we want the snake to delineate our desired object from the image, and at the same time we want a smooth snake boundary.

To formally express the total snake energy as devised by Kass *et al.* we need to parameterize the snaxel coordinates as $(X(s), Y(s))$, where $s \in [0, 1]$. In other words, the contour is parameterized via s, hence we have a parametric contour. Let $I(x, y)$ denote the image intensity at position (x, y) within the image domain. Further, let $f(x, y) = |\nabla I(x, y)|^2$ denote the squared image gradient magnitude. We can now express the total snake energy of Kass *et al.* as follows:

$$E(X, Y) = \frac{1}{2} \int_0^1 \alpha \left(\left| \frac{dX}{ds} \right|^2 + \left| \frac{dY}{ds} \right|^2 \right) + \beta \left(\left| \frac{d^2X}{ds^2} \right|^2 + \left| \frac{d^2Y}{ds^2} \right|^2 \right) ds$$

$$- \int_0^1 f[X(s), Y(s)] ds. \tag{2.1}$$

The first integral embraces the smoothness energy or the so-called "internal energy" of the snake, and the second integral expresses the "external energy" computed over the entire contour. The negative sign in front of the second integral (external energy term) implies that we want to maximize the sum of image gradient magnitude over the entire contour. Note that the internal energy of the snake has two components weighted by non-negative weighting factors α and β. The first of the two internal energy components is known as the stretching energy (likewise, the *tension*) and the second term is called the bending energy (a.k.a. the *rigidity*) for the snake. As the names connote, the stretching energy prevents the snake from getting stretched while the bending energy prevents bending of the snake.

Thanks to the high school calculus lessons given by Dr. Brown at Oakton High School and Dr. Bajani at Howrah Vivekananda Institution, we know that in order to minimize a function of single variable (argument), we simply compute the first derivative of the function with respect to its argument, equate the derivative to zero, and solve the equation. To discover whether the function actually achieves a local minimum value, we need to examine the value of the second derivative of the function at the extremum. If this value is positive then the point is a local minimum. So, on the basis of the lesson of Dr. Brown, one expects that something similar should be done to the snake energy (2.1). Let us first realize that Dr. Brown did not tell us everything we needed to know in life; this high school approach does not work in this case.

The snake energy (2.1) is actually a function of the snake location, which itself is a multi-valued function of s. The energy (2.1) is called a *functional* (a function of a function). So, in the Dr. Brown paradigm, we need to differentiate the snake energy functional with respect to the snake position function. *But how to differentiate a function with respect to a function?* We need an advanced mathematical toolbox called the *calculus of variations* (likewise, *variational calculus*) to perform this minimization [2, 3]. The calculus of variations defines a functional derivative—the derivative of a functional with respect to its argument function(s). So the rest is similar to high school calculus: equate the functional derivative to zero and solve the equation to ferret out the minimizing snake.

In order not to burden the reader uninterested in derivation, we banish the details of computing functional derivatives $\frac{\delta E}{\delta X}$ and $\frac{\delta E}{\delta Y}$ of the energy functional (2.1) to Appendix A. (Note that δ is used here as the functional derivative operator.) First, we equate the functional derivative of (2.1) with respect to the X to zero:

$$\frac{\delta E}{\delta X} = -\alpha \frac{d^2 X}{ds^2} + \beta \frac{d^4 X}{ds^4} - \frac{\partial f}{\partial x} = 0. \tag{2.2}$$

Similarly the functional derivative of (2.1) with respect to Y equated to zero is given as follows:

$$\frac{\delta E}{\delta Y} = -\alpha\frac{d^2 Y}{ds^2} + \beta\frac{d^4 Y}{ds^4} - \frac{\partial f}{\partial y} = 0. \qquad (2.3)$$

Equations (2.2) and (2.3) are known as *Euler equations*. Closed form solutions for X and Y from (2.2) and (2.3) in general cases, where f is any arbitrary potential surface, cannot be computed. We resort to numerical techniques to solve them. One such method is known as the gradient descent method. Consider points A and B in Fig. 2.1(e) again, where the path of the movement of the point from location A to location B is shown. We show the gradient vectors of this potential surface in Fig. 2.2, where we also indicate the point locations A and B. We observe that the gradient vectors are oppositely aligned with the path from A to B. We

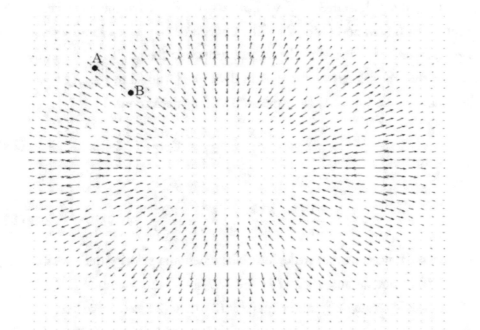

FIGURE 2.2: Gradient vectors of the potential surface in Fig. 2.1(d). Points A and B are also shown here. The direction of gradient is from B to A. Thus the negative of the gradient direction pushes the point A to point B

may intuitively conclude that a point on a potential surface moves in the opposite direction of the gradient of the potential surface. The point stops where the gradient magnitude is zero (for example the point location B). Formally, in gradient descent the velocity of a point (P_x, P_y) is proportional to the negative gradient of the potential surface, $E(x, y)$:

$$\frac{dP_x}{d\tau} \propto -\frac{\partial}{\partial x} E(x, y), \tag{2.4}$$

and

$$\frac{dP_y}{d\tau} \propto -\frac{\partial}{\partial y} E(x, y). \tag{2.5}$$

Note that left sides of Eqs. (2.4) and (2.5) represent the rate of change of the position of the point with respect to time τ, i.e., the velocity of the point. While generating the path from A to B we have actually simulated the movement of the point by Eqs. (2.4) and (2.5). Carefully notice that the velocity of the point at B is zero, where the point stops. In an analogous way, the gradient descent method dictates that the velocity of the snake will be proportional to the negative of the functional gradient of (2.1) (see Appendix A for a derivation):

$$\frac{\partial X}{\partial \tau} = \alpha \frac{d^2 X}{ds^2} - \beta \frac{d^4 X}{ds^4} + \frac{\partial f}{\partial x}, \tag{2.6}$$

and

$$\frac{\partial Y}{\partial \tau} = \alpha \frac{d^2 Y}{ds^2} - \beta \frac{d^4 Y}{ds^4} + \frac{\partial f}{\partial y}. \tag{2.7}$$

As discussed, we crave the snake locations where the velocity (left sides of (2.6) and (2.7)) is zero because then (2.6) and (2.7) fulfill the condition designated by (2.2) and (2.3). A physicist might interpret (2.6) and (2.7) as a force that drives the snake. This resultant force is composed of an *internal force* (stretching and bending terms) and an *external force* (the terms involving the image edge function f). When the resultant force is zero, the snake reaches its equilibrium position and we obtain the desired snake location.

To reach the state of equilibrium, we enact the snake motion governed by Eqs. (2.6) and (2.7), starting from an initial snake location. In order to realize the snake on a discrete grid (*viz.*, a digital image) the continuous contour is approximated by a polygon (in case of a closed contour) or a polyline (for an open contour). As with sampling a function in time, we can increase the number of samples to better approximate the contour. The sampling theorem of Michiganian Claude Shannon can be used to determine the number of samples needed to perfectly reconstruct a continuous and bandlimited (*i.e.*, the frequency of undulations on the contour is bandlimited) contour. These sampled vertices are also commonly referred to as snaxels. In the discrete snake description essentially the continuous parameter, $s \in [0, 1]$ is indexed by $i \in \{0, 1, \ldots, n - 1\}$, where n is the total number of snaxels. A snaxel, located at $(X(s), Y(s))$ on the continuous contour, is denoted by (X_i, Y_i) within the discrete facsimile.

The discrete versions of the Eqs. (2.6) and (2.7) for the ith snaxel are given as follows:

$$\frac{X_i^{\tau+1} - X_i^{\tau}}{\zeta} = \alpha\left(X_{i+1}^{\tau} - 2X_i^{\tau} + X_{i-1}^{\tau}\right) - \beta\left(X_{i+2}^{\tau} - 4X_{i+1}^{\tau} + 6X_i^{\tau} - 4X_{i-1}^{\tau} + X_{i-2}^{\tau}\right) + f_x\left(X_i^{\tau}, Y_i^{\tau}\right), \tag{2.8}$$

and

$$\frac{Y_i^{\tau+1} - Y_i^{\tau}}{\zeta} = \alpha\left(Y_{i+1}^{\tau} - 2Y_i^{\tau} + Y_{i-1}^{\tau}\right) - \beta\left(Y_{i+2}^{\tau} - 4Y_{i+1}^{\tau} + 6Y_i^{\tau} - 4Y_{i-1}^{\tau} + Y_{i-2}^{\tau}\right) + f_y\left(X_i^{\tau}, Y_i^{\tau}\right), \tag{2.9}$$

where the superscripts τ and $\tau + 1$ respectively represent two successive discrete time instants, and the $+$ and $-$ operators appearing in subscripts of Xs and Ys in (2.8) and (2.9) denote modulo n addition and subtraction in order to take care of the wraparound effect for a closed contour. For example, the operation $(n - 1) + 2$ would yield 1, as opposed to $n + 1$. The parameter ζ is the time step that controls the magnitude of steps taken in the discrete updates.

The edge strength terms f_x and f_y are as follows:

$$f_x(x, y) = \frac{\partial}{\partial x} f(x, y) \tag{2.10}$$

and

$$f_y(x, y) = \frac{\partial}{\partial y} f(x, y). \tag{2.11}$$

$(f_x(x, y), f_y(x, y))$, or in shorthand notation (f_x, f_y), forms a vector field over the image domain that acts as an *external force* for the snake. To express (2.8) and (2.9) in matrix-vector notation collectively for all the snaxels, we employ linear algebra:

$$\mathbf{x}^\tau \equiv \left[X_0^\tau, \ldots, X_{n-1}^\tau \right]^\mathrm{T},$$
$$\mathbf{y}^\tau \equiv \left[Y_0^\tau, \ldots, Y_{n-1}^\tau \right]^\mathrm{T},$$
$$\mathbf{f}_x^\tau \equiv \left[f_x(X_0^\tau, Y_0^\tau), \ldots, f_x(X_{n-1}^\tau, Y_{n-1}^\tau) \right]^\mathrm{T},$$

and

$$\mathbf{f}_y^\tau \equiv \left[f_y(X_0^\tau, Y_0^\tau), \ldots, f_y(X_{n-1}^\tau, Y_{n-1}^\tau) \right]^\mathrm{T}.$$

Now (2.8) is rewritten for the entire active contour as

$$\frac{\mathbf{x}^{\tau+1} - \mathbf{x}^\tau}{\zeta} = -A\mathbf{x}^\tau + \mathbf{f}_x^\tau, \tag{2.12}$$

and similarly (2.9) is rewritten as

$$\frac{\mathbf{y}^{\tau+1} - \mathbf{y}^\tau}{\zeta} = -A\mathbf{y}^\tau + \mathbf{f}_y^\tau. \tag{2.13}$$

Here, A is an n-by-n sparse matrix written as

$$A = \begin{bmatrix} c & b & a & & & a & b \\ b & c & b & a & & & a \\ a & b & c & b & a & & \\ & \ddots & \ddots & \ddots & \ddots & \ddots & \\ & & a & b & c & b & a \\ a & & & a & b & c & b \\ b & a & & & a & b & c \end{bmatrix}, \tag{2.14}$$

where in turn a, b, and c are as follows:

$$a = \beta, \quad b = -(4\beta + \alpha), \quad c = 6\beta + 2\alpha. \tag{2.15}$$

To iteratively solve for the snake location (the set of snaxel positions) at time $\tau + 1$, given the position at time τ, we rewrite (2.12) and (2.13) respectively as

$$\mathbf{x}^{\tau+1} = \mathbf{x}^\tau - \zeta(A\mathbf{x}^\tau - \mathbf{f}_x^\tau), \tag{2.16}$$

and

$$\mathbf{y}^{\tau+1} = \mathbf{y}^\tau - \zeta(A\mathbf{y}^\tau - \mathbf{f}_y^\tau). \tag{2.17}$$

Equations (2.16) and (2.17) are known as *explicit* solution techniques [4]. The numerical stability of (2.16) and (2.17) depends on the time step ζ.

An *implicit* procedure [4] that is stable for a wide range of ζ values is followed by rewriting (2.16) and (2.17) as

$$\frac{\mathbf{x}^{\tau+1} - \mathbf{x}^\tau}{\zeta} = -A\mathbf{x}^{\tau+1} + \mathbf{f}_x^\tau, \tag{2.18}$$

and

$$\frac{\mathbf{y}^{\tau+1} - \mathbf{y}^\tau}{\zeta} = -A\mathbf{y}^{\tau+1} + \mathbf{f}_y^\tau. \tag{2.19}$$

Immediately we obtain the following iterative form from (2.18) and (2.19):

$$\mathbf{x}^{\tau+1} = (I_n + \zeta A)^{-1}(\mathbf{x}^\tau + \zeta\mathbf{f}_x^\tau), \tag{2.20}$$

and

$$\mathbf{y}^{\tau+1} = (I_n + \zeta A)^{-1}(\mathbf{y}^\tau + \zeta\mathbf{f}_y^\tau). \tag{2.21}$$

where I_n is the n-by-n identity matrix. It should be noted that the matrix, $(I_n + \zeta A)$ is positive definite (see Appendix B), and so it is invertible. Equations (2.20) and (2.21) are sometimes referred to as the snake evolution equations. Staring from an initial snake location, these evolution equations can be used iteratively to move the snake while tracking a biomedical object, whether it is a cell or the myocardial border or a tumor boundary.

Another way to acquire the snake evolution equations is to represent the snake energy functional approximately in a discrete framework as follows:

$$E(X_0, \ldots, X_{n-1}, Y_0, \ldots, Y_{n-1}) = \frac{1}{2} \sum_{i=0}^{n-1} \alpha(X_{i+1} - X_i)^2 + \alpha(Y_{i+1} - Y_i)^2$$

$$+ \frac{1}{2} \sum_{i=0}^{n-1} \beta(X_{i+1} - 2X_i + X_{i-1})^2 + \beta(Y_{i+1} - 2Y_i + Y_{i-1})^2 - \sum_{i=0}^{n-1} f(X_i, Y_i).$$

$$(2.22)$$

The addition and subtraction operations in the subscript are modulo n addition operations as mentioned in the context of describing Eqs. (2.8) and (2.9). Note that the snake energy functional (2.22) is a function of $2n$ variables $X_0, \ldots, X_{n-1}, Y_0, \ldots, Y_{n-1}$. So if we want to compute the minima of (2.22), we need to take the partial derivatives of (2.22) with respect to these $2n$ variables:

$$\frac{\partial E}{\partial X_i} = -\alpha(X_{i+1} + X_{i-1} - 2X_i) + \beta(X_{i+2} - 4X_{i+1} + 6X_i - 4X_{i-1} + X_{i-2})$$
$$- f_x(X_i, Y_i) \qquad (2.23)$$

and

$$\frac{\partial E}{\partial Y_i} = -\alpha(Y_{i+1} + Y_{i-1} - 2Y_i) + \beta(Y_{i+2} - 4Y_{i+1} + 6Y_i - 4Y_{i-1} + Y_{i-2})$$
$$- f_y(X_i, Y_i), \forall i \in \{0, 1, \ldots, n-1\}. \qquad (2.24)$$

From Eqs. (2.23) and (2.24) we can reach the gradient descent Eqs. (2.8) and (2.9), and derive the snake evolution Eqs. (2.20) and (2.21) in the same way as before. Note that in this latter derivation we altogether bypass the use of calculus of variations, because here we are essentially dealing with functions, not functionals.

Consider again the toy image of Fig. 2.1(a). We have observed in Fig. 2.1 that the snake scrolls down the edge potential surface. (The potential f of Eq. (2.1) in this case is the edge potential surface, $i.e.$, $f(x, y) = |\nabla I(x, y)|^2$.) Let us now look at the snake evolution on the image plane as shown in Fig. 2.3(a). The initial

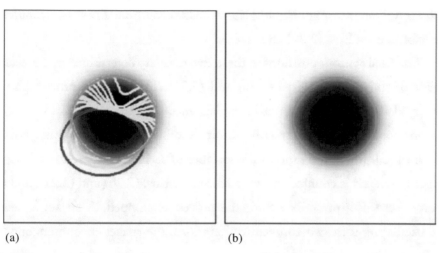

(a)   (b)

FIGURE 2.3: (a) Snake evolution on the circle of Fig. 2.1(a) from the edge potential force. The darker contour represents the initial contour location. The lighter contours represent intermediate contours during the contour evolution

contour shown represents the snake at time zero. The final contour delineating the circle is shown in Fig. 2.3(b).

Another example of snake evolution via edge potential force is given in Fig. 2.4. In Fig. 2.4(a) a leukocyte (white blood cell), as seen from an intravital video

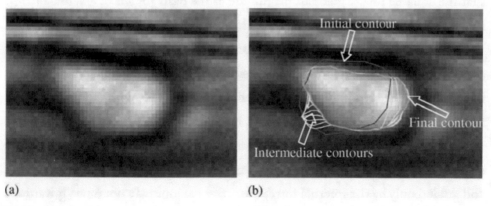

(a)   (b)

FIGURE 2.4: (a) A leukocyte as seen from intravital video microscopy. (b) Capturing the leukocyte with a snake and edge potential force. Initial final and intermediate contours are shown in the figure

microscopy, is shown. Figure 2.4(b) shows snake evolution. These two simulations are performed via Eqs. (2.20) and (2.21).

The final contour positions in these two cases are determined by a prescribed number of iterations for Eqs. (2.20) and (2.21). The iterative process can also be stopped if the average or maximum distance between snaxels in consecutive iterations does not exceed a prescribed limit. Another interesting stopping criterion is given in [5], where the ratio of the number of oscillating snaxels over the total number of snaxels is counted in each iteration of Eqs. (2.20) and (2.21), and once the ratio exceeds a prescribed limit the process is stopped. A snaxel is deemed to be oscillating if in two consecutive iterations, the snaxel encounters opposing directions of the external force. Another very straightforward way for enforcing a stopping criterion is to account for the change in snake energy (2.1). Once the energy reaches a minimum (locally or otherwise) the process can be stopped.

2.3   SNAKE EXTERNAL FORCES

Recall that the overall goal of this chapter is the design of a snake that can be used to capture a moving object in biomedical imaging. It is the external energy that acts as a chemoattractant, drawing the snake to the proper boundary. So far we have seen that the squared image gradient magnitude as the basis for snake external energy. First, we illustrate a basic limitation of this potential energy source for the snake. Let us imagine that the initial active contour is far away (5–10 pixels away, for example) from the coveted edges and residing in a perfectly homogeneous region in an image (as shown in Fig. 2.5). Implementing the snake with the external force field (f_x, f_y) is unsuccessful in this case, because inside the homogeneous region, the image gradient magnitude would be zero and consequently there will be little/no edge force (f_x, f_y) acting on the snake. Unable to "sense" the force due to the edges and guided only by the internal force, the active contour may not move toward the desired edge. In essence, the gradient-based edge potential force field (f_x, f_y) has a limited *capture range*.

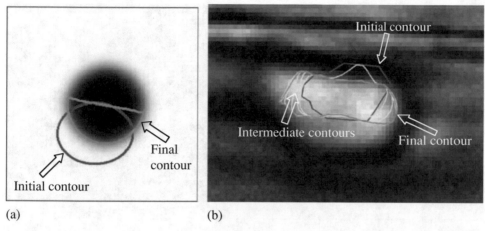

FIGURE 2.5: Illustrations of the failure of the snake method using edge potential force in object boundary delineation. Snake evolutions are shown on (a) synthetic image and on a (b) real image

The distance potential force is a remedy to this limitation for binary images (a binary image has only two intensity levels, 0 and 1). In this case, the distance surface acts as the source of the snake external force. The distance surface (or distance map) $D(x, y)$ is computed from a binary image $I(x, y)$ in the following way:

$$D(x, y) = \min_{(p,q)\in\{(a,b):I(a,b)=1\}} [d(x, y; p, q)], \qquad (2.25)$$

where $d(x, y; p, q)$ is a distance metric between two locations (x, y) and (p, q). For example, d may be Euclidean distance metric:

$$d(x, y; p, q) = \sqrt{(x - p)^2 + (y - q)^2}. \qquad (2.26)$$

The minimum in (2.25) is computed over all locations (p, q) where the binary image intensity is unity. Figure 2.6(a) shows an example distance potential surface. Once we construct $D(x, y)$, we compute the distance potential force field $(-D_x(x, y), -D_y(x, y))$ [Note the appearance of the negative sign: since the

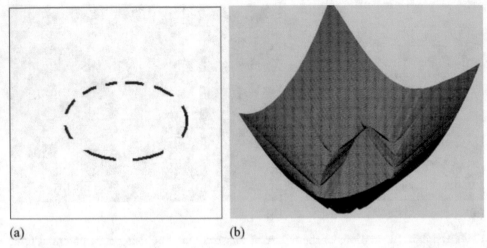

(a)   (b)

FIGURE 2.6: (a) A binary image: an ellipse with a broken boundary. (b) Distance potential surface of the ellipse

negative derivative of the potential is the force], and utilize it in the gradient descent equations:

$$\frac{\partial X}{\partial \tau} = \alpha \frac{d^2 X}{ds^2} - \beta \frac{d^4 X}{ds^4} - \frac{\partial D}{\partial x}, \tag{2.27}$$

and

$$\frac{\partial Y}{\partial \tau} = \alpha \frac{d^2 Y}{ds^2} - \beta \frac{d^4 Y}{ds^4} - \frac{\partial D}{\partial y}. \tag{2.28}$$

Note that Eqs. (2.27) and (2.28) are identical to Eqs. (2.6) and (2.7) except that the edge force (f_x, f_y) is replaced by the distance potential force $(-D_x(x, y), -D_y(x, y))$. The distance potential force is shown in Fig. 2.7(a) and the corresponding snake evolution is illustrated in Fig. 2.7(b).

2.3.1   *The Balloon Force*

A *balloon force* is another type of external force for an active contour [6]. True to the name, the balloon force tries either to inflate or to deflate a closed contour. The balloon force exerts a force that is normal to the active contour (outward or inward). If it is an inflating force then the direction is outward normal; otherwise, it is directed

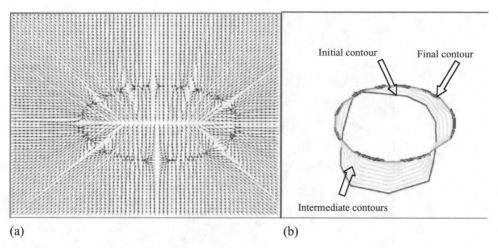

(a)   (b)

FIGURE 2.7: (a) Distance potential force. (b) Snake evolution via distance potential force

to the inward normal. The normal direction to a parameterized contour at a point $(X(s), Y(s))$ is given by the direction of $(-Y_s, X_s)$, $(i.e., \angle(-Y_s, X_s))$. The inward and the outward directions of the normal are determined by the parameterization s: whether s increases or decreases while traveling in a particular direction (for example clockwise) along the curve. So, the balloon force may be quantified by $c_\tau(-Y_s, X_s)$. In other words, the balloon force is proportional to the normal to the curve with a proportionality constant c_τ, which in general may be evolution time dependent. The snake evolution in this case is governed by

$$\frac{\partial X}{\partial \tau} = \alpha \frac{\partial^2 X}{\partial s^2} - \beta \frac{\partial^4 X}{\partial s^2} - c_\tau Y_s, \qquad (2.29)$$

and

$$\frac{\partial Y}{\partial \tau} = \alpha \frac{\partial^2 Y}{\partial s^2} - \beta \frac{\partial^4 Y}{\partial s^2} + c_\tau X_s. \qquad (2.30)$$

The balloon force has gained popularity in medical imaging applications in which an initial contour may be easily placed inside the region to be segmented, as with the lung segmentation application in [7].

In general, we might have more than one type of external force acting on a snake. For example, in Fig. 2.8, we show snake evolution based on two external

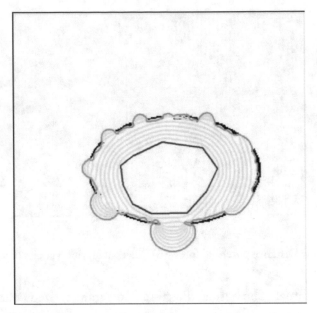

FIGURE 2.8: Snake evolution with balloon force and edge potential force. The contour is seen to "leak" through the edge gaps because of balloon force

forces, an edge force and a balloon force. In such a case the snake evolution equations are given by

$$\frac{\partial X}{\partial \tau} = \alpha \frac{\partial^2 X}{\partial s^2} - \beta \frac{\partial^4 X}{\partial s^2} + \lambda f_x(X, Y) - c_\tau Y_s, \qquad (2.31)$$

and

$$\frac{\partial Y}{\partial \tau} = \alpha \frac{\partial^2 Y}{\partial s^2} - \beta \frac{\partial^4 Y}{\partial s^2} + \lambda f_y(X, Y) + c_\tau X_s. \qquad (2.32)$$

Here λ is a non-negative weight for the edge force.

2.3.2 Gradient Vector Flow

Gradient vector flow (GVF) represents a noteworthy advance in active contour design for biomedical image analysis. In GVF, Xu and Prince [8] construct an external force field $(u(x, y), v(x, y))$ by diffusing the edge force (f_x, f_y), away from edges to the homogeneous regions, at the same time keeping the constructed field as close as possible to the edge force near the edges. They achieve this goal through the minimization of the following energy functional:

$$E_{\text{GVF}}(u, v) = \frac{1}{2} \int \int \mu(u_x^2 + u_y^2 + v_x^2 + v_y^2) + (f_x^2 + f_y^2)((u - f_x)^2 + (v - f_y)^2) dx dy, \tag{2.33}$$

where μ is a non-negative parameter expressing the degree of smoothness of the field (u, v). The interpretation of (2.33) is straightforward—the first integrand keeps the field, (u, v), smooth. This term is quite similar to the solution for the classical Laplace's equation. The second integrand forces the vector field to resemble the initial edge force near the edges (*i.e.*, where the edge force strength is high). Variational minimization of (2.33) results in the following two Euler equations (see Appendix C for derivation):

$$\mu \nabla^2 u - (f_x^2 + f_y^2)(u - f_x) = 0, \tag{2.34}$$

and

$$\mu \nabla^2 v - (f_x^2 + f_y^2)(v - f_y) = 0. \tag{2.35}$$

Solving (2.34) and (2.35) for (u, v) results in gradient vector flow (GVF) that acts as an external force field for the active contour. The derivation details of GVF field are given in Appendix C. Figure 2.9(a) shows the GVF force computed on the circle image of Fig. 2.1(a). Since the GVF vectors exist in homogeneous regions (*i.e.*, where edges are absent) as well, the capture range of the edge force has been effectively increased and the snake correctly captures the circle. In addition to being capable of attracting the active contour from a distance toward the edge, GVF can drag the active contour inside a long concavity (formed by the edges) [8]. Once the GVF force field (u, v) is computed via (2.34) and (2.35), it is utilized in the following snake evolution equations:

$$\frac{\partial X}{\partial \tau} = \alpha \frac{\partial^2 X}{\partial s^2} - \beta \frac{\partial^4 X}{\partial s^2} + u(X, Y), \tag{2.36}$$

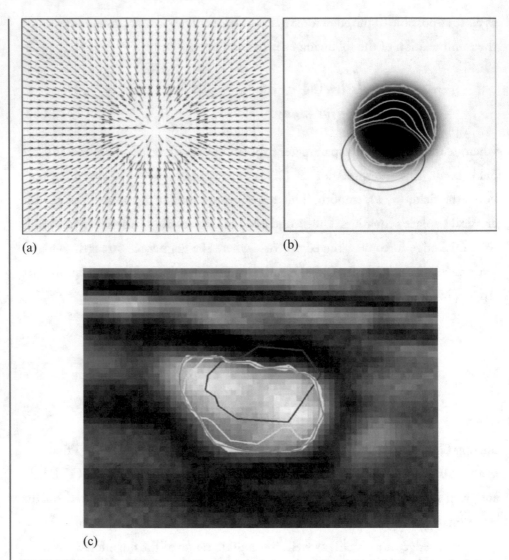

FIGURE 2.9: (a) GVF force field on the circle image of Fig. 2.1(a). (b) Snake evolution via GVF. Same initial contour as Fig. 2.5(a) is used here. (c) GVF snake evolution on leukocyte image. Same initial contour as Fig. 2.5(b) is used here

and

$$\frac{\partial Y}{\partial \tau} = \alpha \frac{\partial^2 Y}{\partial s^2} - \beta \frac{\partial^4 Y}{\partial s^2} + v(X, Y). \tag{2.37}$$

Figure 2.9(b) shows snake evolution via GVF. Notice that the initial contour for this snake is identical to that of Fig. 2.5(a). In contrast to the previous case (using the edge potential force), the snake is able to delineate the circle. Figure 2.9(c) shows GVF snake evolution on the leukocyte image of Fig. 2.4(b). These figures exemplify the increase in capture range provided by GVF. Later in this chapter, we will see that GVF is extremely important in tracking applications, where accurate initialization of the contour is impossible.

2.4 CASE STUDY: TRACKING WITH SNAKES

Snakes are perhaps most widely used in biomedical image segmentation. The treatment of snakes in this book, however, is directed toward the important task of biomedical tracking. We believe the most straightforward way to communicate the snake tracking approach is to use an application-based case study.

Tracking is defined here as the task of following an object through a temporal sequence of images. Tracking in general is composed of two subtasks—object detection and correspondence resolution. Often these two tasks are performed simultaneously, because in general, they can be viewed as interdependent. Correspondence resolution is the problem of identifying a specific object in the current frame that appeared in the previous frame. Of course, when we are tracking only one target and no other targets are present in the video, we do not need any correspondence resolution strategy. Sometimes correspondence resolution is performed through a nearest neighbor assumption. In the nearest neighbor paradigm (used in this case study), the closest detection in space with respect to a previous target position (or predicted target position) is matched to a given target.

Let us now turn our attention to the application of tracking rolling leukocytes observed *in vivo* for this case study. Although the tracking method described in this section is tailored to the rolling leukocyte tracking, this approach may also

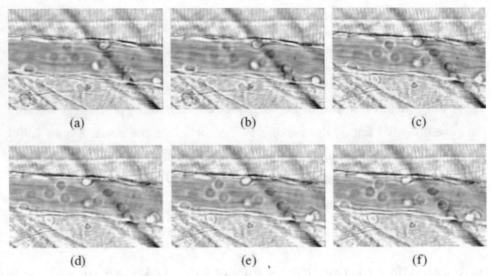

FIGURE 2.10: Six frames from an intravital video sequence showing rolling leukocytes

be applied to tracking cells with well-defined shapes. In Fig. 2.10, a few frames obtained via intravital video microscopy show the motion of rolling leukocytes. Rolling leukocytes are activated leukocytes that move at a much slower speed than the blood flow [9, 10]. With the nearest neighbor assumption, the leukocyte-tracking algorithm with active contours can be described as follows.

Algorithm 2.1

1. *Leukocyte detection.* To initiate this algorithm, a leukocyte on the first frame is detected either manually or automatically, and then active contour evolution is performed to delineate the detected leukocyte.

2. *Tracking.* From the second frame onwards for each frame execute following steps:

   a. *Initial active contour placement.* The final contour delineating the leukocyte from the previous frame is placed over the current video frame.

   b. *Active contour evolution.* Starting from the initial active contour, evolution is performed on the current video frame to delineate the displaced rolling leukocyte.

To delineate a leukocyte with an active contour, we have convinced ourselves that instead of (or in addition to) the smoothness internal energy as described in the work of Kass *et al.*, a shape and size constrained contour is needed. The reason is obvious—a leukocyte is somewhat circular in shape and has a predictable size. Typically a leukocyte has a radius of about 4–7 microns. Thus we require the contour not to deviate much from a circle of a specified radius. This constraint serves as the internal energy for the active contour. We place another constraint on the contour evolution. This constraint comes from the observation about the movement of leukocytes basically follows the blood flow direction. This means the movement of a rolling leukocyte in the direction orthogonal to the blood flow is insignificant. If we align the x-axis in the direction of blood flow, which can be approximated by the venule centerline, then the inter-frame rolling leukocyte movement along the y-axis is limited.

The constrained active contour for leukocyte delineation is expressed as a parametric curve via a reference point (typically the center point), (P, Q), and the polar coordinates $(R(t), t)$; the Cartesian coordinates of the contour points are $(P + R(t)\cos(t), Q + R(t)\sin(t))$. Figure 2.11 depicts such a "radial" active contour. This active contour must be collocated with positions of high gradient magnitude in the image, at the same time the contour should not be deviated significantly from a circular shape of a desired radius. The following energy functional of a shape–size constrained snake serves to delineate

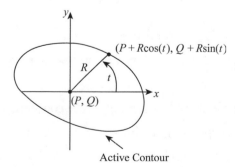

FIGURE 2.11: Radial active contour

leukocytes:

$$E_{r-snake}(P, Q, R) = E_{edge}(P, Q, R) + \mu_{cons} E_{cons}(R) + \mu_{pos} E_{pos}(P, Q, R),$$
(2.38)

where E_{edge} is the external/edge force, E_{cons} is the shape–size constraint, and E_{pos} is the position constraint.

The edge constraint for this application is written as

$$E_{edge}(P, Q, R) = -\frac{1}{L_s} \int_0^{2\pi} w\left[P + R(t)\cos(t), Q + R(t)\sin(t)\right] R(t)dt,$$
(2.39)

where w is a surface that achieves its maxima at edges (such as with the image gradient magnitude), and L_s is the length of the active contour. When the contour is residing on the ridges of this surface w, this energy term is minimized.

The shape–size constraint is expressed as

$$E_{cons}(R) = \frac{1}{2} \int_0^{2\pi} [R(t) - \rho]^2 \, dt,$$
(2.40)

and penalizes deviations of radial distance $R(t)$ from the desired radius, ρ.

The position constraint, for flow along the x-axis, is

$$E_{pos}(P, Q, R) = \frac{1}{2}(Q - P_Y)^2,$$
(2.41)

The position constraint E_{pos} prevents large deviation of the active contour from the estimated direction of leukocyte rolling (P_Y indicates y-coordinate of estimated leukocyte location).

The energy functional (2.38) contains three components with non-negative weights μ_{cons} and μ_{pos} expressing the importance of the respective energy components in the functional. In the subsequent section we discuss how proper values of these parameters can be chosen analytically.

Gradient descent equations for minimizing (2.38) are obtained by variational calculus as follows (see Appendix D):

$$\frac{\partial P}{\partial \tau} = \overline{w}_x, \tag{2.42}$$

where τ here is pseudotime,

$$\frac{\partial Q}{\partial \tau} = \overline{w}_y - \mu_{\text{pos}}(Q - P_y), \tag{2.43}$$

and

$$\frac{\partial R(t)}{\partial \tau} = \frac{1}{L_s}\left[w + R(t)\frac{\partial w}{\partial x}\cos(t) + R(t)\frac{\partial w}{\partial y}\sin(t) - \overline{w}\right] - \mu_{\text{cons}}\left[R(t) - \rho\right], \tag{2.44}$$

where

$$\overline{w} = \frac{1}{L_s}\int_0^{2\pi} w\left[P + R(t)\cos(t),\ Q + R(t)\sin(t)\right] R(t)dt, \tag{2.45}$$

$$\overline{w}_x = \frac{1}{L_s}\int_0^{2\pi} \frac{\partial w}{\partial x}\left[P + R(t)\cos(t),\ Q + R(t)\sin(t)\right] R(t)dt, \tag{2.46}$$

and

$$\overline{w}_y = \frac{1}{L_s}\int_0^{2\pi} \frac{\partial w}{\partial y}\left[P + R(t)\cos(t),\ Q + R(t)\sin(t)\right] R(t)dt. \tag{2.47}$$

In steps (2a) and (2b) of Algorithm 2.1, we iteratively employ Eqs. (2.42)–(2.44) starting from an initial contour to obtain a new active contour configuration that locally minimizes (2.38) and delineates a leukocyte in the process. In the next section we discuss a suitable construction process for the edge potential surface w that facilitates leukocyte delineation via shape–size constrained active contours.

2.4.1 External Force for Cell Tracking Case Study

An edge potential force derived from the image gradient is limited to the local proximity of the edges. Tracking with a method such as Algorithm 2.1 would be successful with such an external force if the frame-to-frame displacement of the target were not large. However, if the displacements exceed several pixels, the target will be lost. Consider a rolling leukocyte observed at the standard video capture rate of 30 frames per second and a resolution of three pixels per micron; if the leukocyte velocity exceeds 60 microns/s, the displacement will be on the order of 6 pixels per frame. In such a case, it is unlikely that the snake will "see" the boundary if the guiding force is based solely on the intensity gradient. One remedy is to increase the frame rate. However, then computational expense increases because there are more frames to process. Another remedy is to have a slightly modified version of Algorithm 2.1 where we advance the contour from the previous frame in the direction of blood flow so that the advanced contour becomes close to the leukocyte boundary on the current frame. At this point, the knee-jerk question would be— *how far should the contour be advanced*? One approach could involve learning the leukocyte movement pattern; then predicting this advancement using the Kalman filter or some other predictor and estimator. However rolling leukocytes exhibit various mechanisms of movement while rolling along the microvessel wall—they may halt briefly, then make a sudden jump, and then continue steadily at a constant velocity. It is also not uncommon that a leukocyte exhibits combinations of some or even all of these movement patterns. Therefore, prediction of movement with a constant velocity model is likely to be unsuccessful.

2.4.2   Motion Gradient Vector Flow

What if we used a gradient vector flow field that was biased in the known direction of motion? We can design an external force field for the shape, size, and position constrained snake taking into account cell movement direction so that a lagging initial contour will be drawn toward the cell edge on the current frame. The external force should also be able to handle the case where the frame-to-frame leukocyte displacement is small, or nearly zero. As we have already stated while describing the position constraint, the cell motion direction more or less follows the blood flow. Therefore this direction of cell movement can be estimated *a priori* when extracting the microvessel boundary. (See Section 2.5 for a discussion of microvessel boundary detection.)

Motion gradient vector flow (MGVF) is an external force that can be utilized in Algorithm 2.1 to track a moving object with a frame-to-frame displacement that is less than its diameter. We first want to represent the gradient magnitude surface $f = |\nabla I|$ by a surface w; then gradient of this surface (*i.e.*, ∇w) will serve as the external force field for the snake. We may argue that this surface should have two properties: (a) the slope of this surface inside a cell should be such that a contour from the previous frame, even with minimal overlap with the cell, will be dragged by the surface slope toward leukocyte delineation; (b) once the contour reaches the leukocyte edge, it should cease movement (*i.e.*, it achieves a state of equilibrium). Minimizing the following energy functional we create such a surface w [9].

$$E_{\mathrm{MGVF}}(w) = \frac{1}{2} \iint \{\mu H_\varepsilon[\nabla w \cdot (v^x, v^y)]|\nabla w|^2 + f(w - f)^2\}dxdy, \qquad (2.48)$$

where (v^x, v^y) is the known blood flow direction, and H_ε is a regularized (by the positive parameter ε) Heaviside function that is a continuously differentiable approximation to the unit step function:

$$H_\varepsilon(z) = \frac{1}{2}\left(1 + \frac{2}{\pi}\tan^{-1}\left(\frac{z}{\varepsilon}\right)\right). \qquad (2.49)$$

Because of the presence of the Heaviside function in (2.48), the diffusion of w is maximized when the Heaviside function achieves a value of unity, and it is minimized when the Heaviside function is zero. In other words, when the vectors ∇w (∇w serves as the external force in MGVF) and (v^x, v^y) are aligned with each other, the diffusion is maximal. Also because of the $f(w - f)^2$ term in the integral of (2.48), the surface w remains close to f whenever the value of f is high. The parameter μ is a non-negative constant controlling the contribution of the first (diffusivity) term.

Applying the variational principles along with a bag of minor mathematical tricks (see Appendix E) for the minimization of (2.48), we obtain a gradient descent equation that can be used to derive the motion gradient vector flow field:

$$\frac{\partial w}{\partial \tau} = \mu \, \text{div}\{H_\varepsilon[\nabla w \cdot (v^x, v^y)]\nabla w\} - f(w - f). \qquad (2.50)$$

The diffusion mechanism is clearly understood from the "div" (divergence) term in Eq. (2.50). It is an anisotropic diffusion with a diffusion coefficient $H_\varepsilon(\nabla w \cdot (v^x, v^y))$ that encourages diffusion where ∇w and (v^x, v^y) form acute angles and discourages diffusion where they form obtuse angles. Once (2.50) is applied, the solution surface w serves as the negative of the potential for the snake (*viz.*, ∇w), acting as the external force for the snake. We refer to this force field ∇w as the motion gradient vector flow force.

2.4.3   Computation of Motion Gradient Vector Flow Field

To obtain a solution to (2.50), we follow an eight-neighborhood system on the discrete Cartesian image domain and utilize a Jacobian solution procedure as was used in solving the traditional anisotropic diffusion equation [11].

$$w_{i,j}^{\tau+1} = w_{i,j}^\tau + \frac{\mu}{\lambda} \sum_{l=-1}^{1} \sum_{m=-1}^{1} H_\varepsilon\big[(lv^y + mv^x)\big(w_{i+l,j+m}^\tau - w_{i,j}^\tau\big)\big]\big(w_{i+l,j+m}^\tau - w_{i,j}^\tau\big)$$
$$- \frac{1}{\lambda} f_{i,j}\big(w_{i,j}^\tau - f_{i,j}\big). \qquad (2.51)$$

Here $w_{i,j}^0 = f_{i,j}$, $w_{i,j}^\tau$, and $f_{i,j}$ respectively denote the value of the surface w and the edge-map f at the (i, j)th location in the discrete domain, τ denotes the iteration number, and λ denotes inverse of the time-step. The following proposition illustrates the convergence conditions and the speed of convergence for (2.51).

Proposition 1.   The numerical implementation given by (2.51) is convergent, and the rate of convergence is that of a geometric series of common ratio ϖ, provided

(i).   the edge-map f is normalized such that

$$0 < \varpi \leq f_{i,j} \leq 1, \quad \forall i, j, \qquad (2.52)$$

and

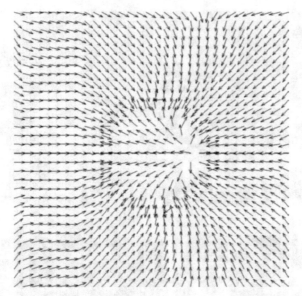

FIGURE 2.12: MGVF force field on the synthetic circle image. The direction of motion (v^x, v^y) is from right to left here

(ii).   we select the multiplicative inverse of the time-step as

$$\lambda \geq 1 + 8\mu. \qquad (2.53)$$

Proof. See Appendix F.

Let us illustrate the efficacy of MGVF through a couple of examples. Figure 2.12 shows MGVF field ∇w for the circle image of Fig. 2.1(a). We have assumed $v^x = -1$, $v^y = 0$ (*i.e.*, the circle is moving in the negative x-direction). The vector field MGVF is quite different from the GVF vector field for the circle shown in Fig. 2.9(a). A lagging active contour can now be attracted to the circle edges.

The next set of figures illustrates this point for a leukocyte-tracking example. Figure 2.13(a) shows the leukocyte and Fig. 2.13(b) shows the corresponding MGVF force field for $v^x = -1$, $v^y = 0$. Figures 2.13(c)–2.13(d) shows leukocyte delineation by implementing a snake via gradient descent Eqs. (2.42)–(2.44) in conjunction with the computed w via (2.51).

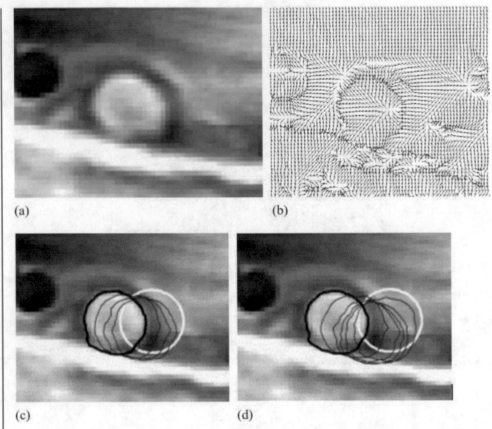

FIGURE 2.13: (a) Leukocyte and (b) corresponding MGVF field. The direction of motion (v^x, v^y) is from right to left here. (c) and (d) Snake evolution via MGVF and shape–size–position constrained snake. White, gray, and dark contours respectively represent initial, intermediate, and final snakes.

2.5   CHOOSING PARAMETER VALUES

One of the aspects of snake-based tracking that is apt to a knob-tweaking solution is the question of weighting parameter selection. The *minimax* criterion provides an analytical way to determine the parameters such as μ_{cons} and μ_{pos} of (2.38) involved in multi-component energy functionals [12]. To apply the minimax criterion, the individual energy components must be non-negative and the energy functional must be a convex combination of the constituent energy components. The combination may be linear, quadratic or cubic, and so on. For example, the weighting parameters in the quadratic combination are non-negative and when squared they add up to unity. To conform to these requirements of the minimax criterion, we adopt the energy functional

$$E_{\text{nonlin-rsnk}}(P, Q, R, \mu_{\text{cons}}, \mu_{\text{pos}}) = \sqrt{(1 - \mu_{\text{cons}}^2 - \mu_{\text{pos}}^2)}[1 + E_{\text{edge}}(P, Q, R)]$$
$$+ \mu_{\text{cons}} E_{\text{cons}}(R) + \mu_{\text{pos}} E_{\text{pos}}(P, Q, R),$$

$$(2.54)$$

where E_{edge}, E_{cons}, and E_{pos} are defined by (2.39), (2.40), and (2.41) respectively. Note that the value of E_{edge} defined via (2.39) lies between -1 and 0; therefore, in order to make this component non-negative we add 1 to E_{edge} in (2.54).

The minimax principle minimizes (2.54) as follows:

$$(P^*, Q^*, R^*) = \arg\min_{P,Q,R}[\arg\max_{\mu_{\text{cons}},\mu_{\text{pos}}} E_{\text{nonlin-rsnk}}(P, Q, R, \mu_{\text{cons}}, \mu_{\text{pos}})]$$
$$= \arg\max_{\mu_{\text{cons}},\mu_{\text{pos}}}[\arg\min_{P,Q,R} E_{\text{nonlin-rsnk}}(P, Q, R, \mu_{\text{cons}}, \mu_{\text{pos}})]$$
$$= \arg\max_{\mu_{\text{cons}},\mu_{\text{pos}}}[E_{\text{nonlin-rsnk}}^*(\mu_{\text{cons}}, \mu_{\text{pos}})], \qquad (2.55)$$

where

$$E_{\text{nonlin-rsnk}}^*(\mu_{\text{cons}}, \mu_{\text{pos}}) = \min_{P,Q,R} E_{\text{nonlin-rsnk}}(P, Q, R, \mu_{\text{cons}}, \mu_{\text{pos}}). \qquad (2.56)$$

We know that the function $E_{\text{nonlin-rsnk}}^*$ is concave up [12]. In consequence, the parameter value set, $(\mu_{\text{cons}}^*, \mu_{\text{pos}}^*)$, corresponding to the minimax criterion, is now determined uniquely by (see [12])

$$(\mu_{\text{cons}}^*, \mu_{\text{pos}}^*) = \arg\max_{\mu_{\text{cons}},\mu_{\text{pos}}} E_{\text{nonlin-rsnk}}^*(\mu_{\text{cons}}, \mu_{\text{pos}}). \qquad (2.57)$$

Unfortunately, multiple minimization computations for the energy functional (2.54) are required to determine the required parameter value set $(\mu_{\text{cons}}^*, \mu_{\text{pos}}^*)$. Since the function (2.56) is concave up, we can set up a simple "steepest ascent" type search method to crawl up the top of the surface $E_{\text{nonlin-rsnk}}^*$:

Algorithm 2.4.1

1. Start with initial values $\mu^*_{\text{cons}} = \mu^0_{\text{cons}}$ and $\mu^*_{\text{pos}} = \mu^0_{\text{pos}}$.

2. Choose a step size $h > 0$.

3. for $n = 1$:Max_Iterations

$$(\mu^n_{\text{cons}}, \mu^n_{\text{pos}}) = \underset{\substack{\mu_1 \in \{\mu^n_{\text{cons}}-h, \mu^n_{\text{cons}}, \mu^n_{\text{cons}}+h\} \\ \mu_2 \in \{\mu^n_{\text{pos}}-h, \mu^n_{\text{pos}}, \mu^n_{\text{pos}}+h\}}}{\arg\max} \left[E^*_{\text{nonlin-rsnk}}(\mu_1, \mu_2) \right.$$
$$\left. - E^*_{\text{nonlin-rsnk}}(\mu^{n-1}_{\text{cons}}, \mu^{n-1}_{\text{pos}}) \right]$$

4. Output: $\mu^*_{\text{cons}} = \mu^n_{\text{cons}}$ and $\mu^*_{\text{pos}} = \mu^n_{\text{pos}}$.

The loop in Algorithm 2.4.1 performs hill-climbing on the surface $E^*_{\text{nonlin-rsnk}}$. Instead of a maximum iteration value for the loop, the iterative process can also be terminated when the change between the values $(\mu^n_{\text{cons}}, \mu^n_{\text{pos}})$ and $(\mu^{n-1}_{\text{cons}}, \mu^{n-1}_{\text{pos}})$ becomes insignificant. Note that once $(\mu^*_{\text{cons}}, \mu^*_{\text{pos}})$ is found via Algorithm 2.4.1, we achieve the desired minimax solution for P, Q, and R from (2.54).

Figure 2.14(a) shows the surface plot of $E^*_{\text{nonlin-rsnk}}$ vs.. $(\mu_{\text{cons}}, \mu_{\text{pos}})$ for a sample leukocyte image. Figure 2.14(b) shows a plot of solution quality vs. $(\mu_{\text{cons}}, \mu_{\text{pos}})$. To find out the solution quality value for a set of value $(\mu_{\text{cons}}, \mu_{\text{pos}})$, we first compute (2.56) and then calculate the Pratt figure of merit (FOM) for the solution [13]:

$$\text{FOM} = \frac{1}{\max(N_d, N_i)} \sum_{n=1}^{N_d} \frac{1}{1 + \alpha d_n^2}, \tag{2.58}$$

where N_d and N_i are respectively the detected and the actual (true) number of edge points on an image, d_n is the distance between the nth true edge point from its nearest detected edge point, and α is a weighting parameter ($\alpha = 1/9$ here). Pratt's FOM is bound between 0 and 1, with unity representing the perfect segmentation. Note from Figs. 2.14(a) and 2.14(b) that the minimax parameter value approximately corresponds to highest solution quality.

On a side note, we introduce the Pratt figure of merit here as a quantitative measure of segmentation success. We believe that there is a striking absence of established quantitative measures of success in image analysis tasks. We feel that

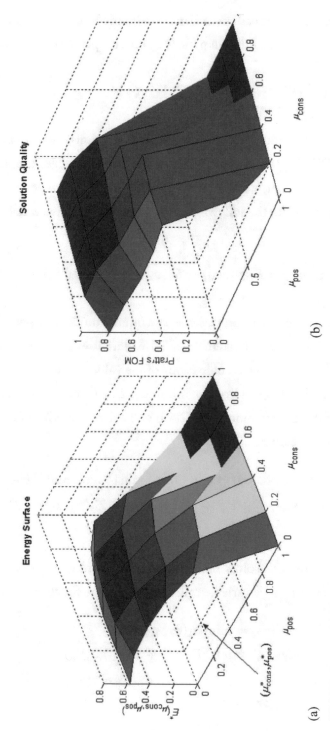

FIGURE 2.14: (a) Energy values *vs.* weighting parameters. (b) Solution quality *vs.* weighting parameter values

the establishment of such measures is critical to biomedical image analysis, where the results may affect the health of a patient or the acceptance of a novel drug.

2.6   DYNAMIC PROGRAMMING FOR SNAKE EVOLUTION

As an alternative to gradient descent based snake solutions, we explore dynamic programming (DP) [14]. DP is effective when not all variables are interrelated simultaneously in the energy functional that we seek to minimize. To make this point clear, let us consider the following energy functional of five variables, v_1, v_2, v_3, v_4, and v_5:

$$E(v_1, v_2, v_3, v_4, v_5) = E_1(v_1, v_2) + E_2(v_2, v_3) + E_3(v_3, v_4) + E_4(v_4, v_5).$$

(2.59)

Let us assume that each decision variable v_i can take on only m possible values. So if we want to minimize E via an exhaustive search we first need to compute m^5 values of E for all possible combinations of the values of the five variables and then pick the combination yielding the minimum E. In contrast, DP exploits the additive form of the energy functional (2.59) and solves the minimization in five sequential stages as follows:

$$\min_{v_1, v_2, v_3, v_4, v_5} E(v_1, v_2, v_3, v_4, v_5) = \min_{v_5} \Big\{ \min_{v_4} \Big[\min_{v_3} \Big(\min_{v_2} \Big\{ \min_{v_1} [E_1(v_1, v_2)] + E_2(v_2, v_3) \Big\} + E_3(v_3, v_4) \Big) + E_4(v_4, v_5) \Big] \Big\}.$$

(2.60)

In other words, DP solves the minimization problem by generating a sequence of functions of single variable called optimal value functions:

$$
\begin{aligned}
D_1(v_2) &= \min_{v_1}[E_1(v_1, v_2)], \\
D_2(v_3) &= \min_{v_2}[D_1(v_2) + E_2(v_2, v_3)], \\
D_3(v_4) &= \min_{v_3}[D_2(v_3) + E_3(v_3, v_4)], \\
D_4(v_5) &= \min_{v_4}[D_3(v_4) + E_4(v_4, v_5)], \\
D_5 &= \min_{v_5}[D_4(v_5)].
\end{aligned}
$$

(2.61)

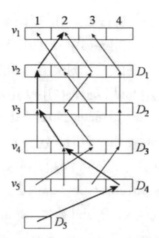

FIGURE 2.15: Dynamic programming: optimal value functions in an example situation. The optimal path is shown with boldface arrows. Four values for each of the five variables $(v_1, v_2, \ldots, v_5)$ are possible. Consequently there are four possible values for each of the five optimal value functions $(D_1, D_2, \ldots, D_5)$. See text for description

Since $D_1(v_2)$ is a function of v_2, which can assume m values, $D_1(v_2)$ can be represented by an array of length m. To compute $D_1(v_2)$, for each value of v_2, we find a value of v_1 that yields the minimum value for $E_1(v_1, v_2)$. This minimum value is assigned to the corresponding element of the $D_1(v_2)$ array. Each element of the $D_1(v_2)$ array also points to the corresponding value of v_1 which has yielded the minimum $E_1(v_1, v_2)$.

For the sake of illustration, let $m = 4$. In Fig. 2.15 we show only the pointers from the $D_1(v_2)$ array pointing to the index array: 1 through 4. By pointing the second element of $D_1(v_2)$ array to value 1, we indicate that out of the four values $E_1[v_1(1), v_2(2)]$, $E_1[v_1(2), v_2(2)]$, $E_1[v_1(3), v_2(2)]$, and $E_1[v_1(4), v_2(2)]$, the minimum energy value is $E_1[v_1(1), v_2(2)]$. Similarly, to indicate that $E_1[v_1(2), v_2(3)]$ is the minimum of the four values $E_1[v_1(1), v_2(3)]$, $E_1[v_1(2), v_2(3)]$, $E_1[v_1(3), v_2(3)]$, and $E_1[v_1(4), v_2(3)]$, the third element of array $D_1(v_2)$ points to value 2. To compute $D_2(v_3)$, for each value of v_3 we find the value of v_2 that yields the minimum $[D_1(v_2) + E_2(v_2, v_3)]$. This minimum value is assigned to $D_2(v_3)$ along with a pointer to the corresponding element of the $D_1(v_2)$ array. In

Fig. 2.15, we show the pointers from $D_2(v_3)$. As an example, we point $D_2[v_3(2)]$ to $D_1[v_2(3)]$ to indicate that $\{D_1[v_2(3)]+E_2[v_2(3),v_3(2)]\}$ is the minimum of the four values: $\{D_1[v_2(1)]+E_2[v_2(1),\ v_3(2)]\}$, $\{D_1[v_2(2)]+E_2[v_2(2),\ v_3(2)]\}$, $\{D_1[v_2(3)]+E_2[v_2(3),\ v_3(2)]\}$ and $\{D_1[v_2(4)]+E_2[v_2(4),\ v_3(2)]\}$. Arrays $D_3(v_4)$ and $D_4(v_5)$ are computed in like manner. Note that D_5 is a single element array pointing to $D_4(v_5(4))$ indicating that $D_4[v_5(4)]$ is the minimum of the four values: $D_4[v_5(1)]$, $D_4[v_5(2)]$, $D_4[v_5(3)]$, and $D_4[v_5(4)]$. Now we can trace the path (shown in bold in Fig. 2.15) starting at D_5. Figure 2.15 shows a set of links in bold that depicts $v_1(2)$, $v_2(1)$, $v_3(1)$, $v_4(2)$, and $v_5(4)$ as the values of the five variables resulting in the minimum energy functional E of (2.59).

Let us now analyze the computational complexity of the minimization of (2.59) via DP. In computing $D_1(v_2)$ we evaluate the objective function component $E_1(v_1,v_2)$ m^2 times. Similarly in computing $D_2(v_3)$, $D_3(v_4)$, and $D_4(v_5)$ it takes m^2 such evaluations of the minimands (the objective functions to be minimized) in each case. Finally in order to compute D_5, it takes m evaluations of minimands. Thus, in total, DP takes $4m^2 + m$ evaluations, as opposed to m^5 evaluations resulting from the exhaustive search method. Likewise, if the objective function E in (2.59) had n constituent function components in total, then DP would take $(n-1)m^2 + m$ evaluations. So DP has a computational complexity of $O(nm^2)$ for minimizing (2.59) as opposed to $O(m^n)$ for the exhaustive search.

Let us now consider an objective function with the following form:

$$E(v_1, v_2, v_3, v_4, v_5) = E_1(v_1, v_2) + E_2(v_2, v_3) + E_3(v_3, v_4) + E_4(v_4, v_5)$$
$$+E_5(v_5, v_1). \tag{2.62}$$

In this case DP generates the following optimal value functions of two variables:

$$D_1(v_1, v_3) = \min_{v_2}[E_1(v_1, v_2) + E_2(v_2, v_3)],$$
$$D_2(v_1, v_4) = \min_{v_3}[D_1(v_1, v_3) + E_3(v_3, v_4)],$$
$$D_3(v_1, v_5) = \min_{v_4}[D_2(v_1, v_4) + E_4(v_4, v_5)],$$
$$D_4 = \min_{v_5, v_1}[D_3(v_1, v_5) + E_5(v_5, v_1)]. \tag{2.63}$$

Note that the computational complexity of DP, to minimize objective functions with the form of (2.61), is $O(nm^3)$ where n is the total number of constituent energy terms. Sometimes the objective function is composed of constituent functions of three variables. An example is as follows:

$$E(v_1, v_2, v_3, v_4, v_5) = E_1(v_1, v_2, v_3) + E_2(v_2, v_3, v_4) + E_3(v_3, v_4, v_5).$$
$$(2.64)$$

DP then generates the following optimal value functions of two variables:

$$D_1(v_2, v_3) = \min_{v_1}[E_1(v_1, v_2, v_3)],$$
$$D_2(v_3, v_4) = \min_{v_2}[D_1(v_2, v_3) + E_2(v_2, v_3, v_4)],$$
$$D_3(v_4, v_5) = \min_{v_3}[D_2(v_3, v_4) + E_3(v_3, v_4, v_5)],$$
$$D_4 = \min_{v_4, v_5} D_3(v_4, v_5).$$

The energy functional of the form (2.64) is referred to as energy functional of second order interaction terms, whereas (2.59) is known as energy functional of first order interactions. It is not difficult to see that DP minimizes objective functions with second order interactions in $O(nm^3)$, where n is the total number of terms in the energy functional. *But how do we apply DP to tracking with snakes?*

For the sake of simplicity, we consider DP snake computation where the snake energy functional has only first order interaction terms as follows:

$$E(X_0, \ldots, X_{n-1}, Y_0, \ldots, Y_{n-1}) = \frac{1}{2}\sum_{i=0}^{n}\alpha(X_{i+1} - X_i)^2 + \alpha(Y_{i+1} - Y_i)^2$$
$$-\sum_{i=0}^{n}f(X_i, Y_i). \qquad (2.65)$$

If the snake is a closed contour, then the functional (2.64) is similar to (2.61). In this case, DP leads to the following optimal value functions:

$$D_0(X_0, Y_0, X_2, Y_2) = \min_{X_1, Y_1}[E_0(X_0, Y_0, X_1, Y_1) + E_1(X_1, Y_1, X_2, Y_2)],$$
$$D_1(X_0, Y_0, X_3, Y_3) = \min_{X_2, Y_2}[D_0(X_0, Y_0, X_2, Y_2) + E_2(X_2, Y_2, X_3, Y_3)],$$
$$\vdots$$

$$D_{n-2}(X_0, Y_0, X_n, Y_n) = \min_{X_{n-1}, Y_{n-1}} [D_{n-3}(X_0, Y_0, X_{n-1}, Y_{n-1})$$
$$+ E_{n-1}(X_{n-1}, Y_{n-1}, X_n, Y_n)],$$
$$D_{n-1} = \min_{X_0, Y_0, X_n, Y_n} [D_{n-2}(X_0, Y_0, X_n, Y_n)$$
$$+ E_n(X_n, Y_n, X_0, Y_0)]. \tag{2.66}$$

where

$$E_i(X_i, Y_i, X_{i+1}, Y_{i+1}) = \tfrac{1}{2}\alpha(X_{i+1} - X_i)^2 + \tfrac{1}{2}\alpha(Y_{i+1} - Y_i)^2 - f(X_i, Y_i),$$
$$i = 0, \ldots, n. \tag{2.67}$$

The optimal snake location is found by tracing the variable values from D_{n-1} to D_0. Note that in the snake computation the energy functional has to be minimized iteratively until the energy functional value reaches a local minimum. Therefore the optimal value functions (2.65) are generated in each iteration of snake computation until no further change in snake position occurs. A snake energy functional with second order interactions can be similarly handled via DP. However, the computational complexity will be slightly higher.

Figure 2.16 shows DP snake evolutions used to compute the microvessel boundary from an *in vivo* microscopy observation.

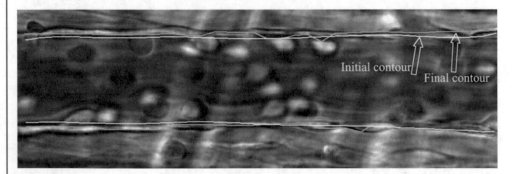

FIGURE 2.16: *In vivo* image showing venule boundaries. Snake evolution by dynamic programming (DP) is used to compute venule boundaries. Initial and final snakes are shown by arrows. This is an example of open-ended snake evolution

2.7   CONCLUSIONS

Our first tool for tracking the boundaries of biological objects is the traditional snake, powered by the expeditious mechanism of gradient descent. Where optimization variables can be sequentially treated, dynamic programming may provide a superior solution.

Several key concepts such as internal energy, external energy, and the snake force are necessary in mastering the snake. This chapter also puts forth a toolbox of snake external forces including the state-of-the-art gradient vector flow. For objects in motion, we prescribe a specially designed external force, motion gradient vector flow. The chapter provides Algorithm 2.1, which is a simple yet powerful method that can be exploited in many object-tracking applications. We show how the tracking algorithm can be tailored to certain biological objects using a shape–size–position constrained snake. Finally, a potential Achilles' heel in snake tracking—how to choose the weighting parameter values in the energy functional—is tackled by way of the minimax method.

CHAPTER 3

Bayesian Tracking and the Kalman Filter

"It is now known that modeling randomness in the real world by means of, or as a consequence of, probability theory is not only naïve but unscientific."

—Rudolf Kalman

"I'm a great believer in luck, and I find the harder I work, the more luck I have."

—Thomas Jefferson

3.1 OVERVIEW

The methods limned in Chapter 2 are deterministic. The difficulty with the deterministic methods, is that sometimes, as Martin Sexton put it, we do not know the destination, especially with tracking biomedical objects in motion. To aid us in "guessing" where the object of interest is going to move, or to divine where the object has indeed moved, we use our knowledge of probability and stochastic processes. In this context, we first discuss the sequential Bayesian framework for target tracking. The tracking workhorse for the past 30 years has been the Kalman filter. We describe the efficacy of the Kalman filter in tracking biomedical objects and detail a special case that is widely utilized in tracking—the alpha–beta filter. We also describe the extended Kalman filter as a tool applicable to nonlinear dynamics.

It is our goal in discussing the Kalman filter not to replicate the fine derivations and proofs of optimality found in existing texts, but to bring the Kalman filter to the biomedical tracking application.

3.2 SEQUENTIAL BAYESIAN FILTERING

In this section we first outline the basic theory of sequential Bayesian filtering for tracking. We illustrate how the basic equations lead one in the labyrinth from the Kalman filter to the particle filter (covered in Chapter 4). Throughout, we use X_t and Z_t to represent the state and the observation (measurement) respectively at the current time t, for a tracking system. Typically X_t denotes position, velocity, and other similar properties of a target that we need to estimate in order to track. In the biomedical world, X_t might contain information on shape, intensity profile, and salient high information points. The state X_t is "hidden" in the sense that we cannot directly observe or measure it. We can only directly observe the process Z_t, and we have some mathematical relationship that links X_t and Z_t. For example, Z_t may be derived from image intensity values, or may even be an observed target position (possibly obtained by template matching). In the case where Z_t denotes target position—we say that it is only the noisy measured/observed target location, and this measurement is not the "true target position" (which is couched in X_t). In such a case, the tracking problem becomes one of estimating true target position X_t from the noisy observation Z_t.

Sequential Bayesian tracking formulation assumes that we know the following three probability densities (or probability mass functions in the case of discrete variables):

1. $p(X_0)$,

2. $p(X_t|X_{t-1})$, for $t \geq 1$, and

3. $p(Z_t|X_t)$, for $t \geq 1$,

where $p(X_0)$ represents the initial distribution of the system state. The state transition probability, and $p(X_t|X_{t-1})$ represents a Markovian state model. The measurement probability, $p(Z_t|X_t)$, represents the dependence of the observation on the actual target state. The density $p(X_t|X_{t-1})$ is Markovian because we assume that the probability of the current state depends only on the immediate past state (i.e., $p(X_t|X_{t-1}) = p(X_t|X_1, X_2, \ldots, X_{t-1})$). The measurement density $p(Z_t|X_t)$ is called conditionally independent because the observation Z_t depends only upon the current state X_t and is independent of other random variables. We remind you that we can directly measure the observations (Z_t), but we do not observe or measure the state (X_t) directly. Thus, the tracking problem is manifested as an estimation of the current target state X_t given all the observations Z_1 through Z_t available at the current time t.

Before delving into solving this estimation problem in the sequential Bayesian setting, we illustrate a toy example of the system we just described.

Let us consider the movement of a bright target of radius r on a 2D image plane with a dark background—a bright "blob" on a dark background. Movements are time-stamped as 0, 1, 2, ..., t, where t is the current time. The initial center coordinates of the target are fixed and specified by x_0; the center coordinates of the blob at current time stamp t are given by X_t. Let us assume for the sake of illustration that we do not have a mechanism to observe the target center coordinates; rather, we only know its average intensity and the standard deviation of the intensity within it at each time t. Thus the observation Z_t in this case is the target intensity at time t. The requisite densities are

$$p(X_0) = \delta(X_0 - x_0), \tag{3.1}$$

$$p(X_t|X_{t-1}) \propto \exp\{b \, \cos[\arg(X_t - X_{t-1})]\} \exp(-|X_t - X_{t-1}|^2/2\sigma^2), \text{ for } t \geq 1, \tag{3.2}$$

and

$$p(Z_t|X_t) \propto \exp\left\{-|Z_t - g(X_t)|^2/2\tau^2\right\}, \quad \text{for } t \geq 1, \tag{3.3}$$

where $g(X_t)$ computes the average intensity inside a circle of radius r centered at X_t, σ, and τ are respectively known standard deviations. The concentration parameter b for the von Mises distribution can be approximately described as a truncated Gaussian with support $[0, 2\pi]$ and variance $1/b$. Note that the motion model (3.2) dictates that the velocity direction of the target is distributed as a von Mises distribution with zero mean, and the speed (velocity magnitude) is distributed as a zero mean Gaussian. Equation (3.1) in this case represents a point mass concentrated at x_0, where δ is the Kroneker delta function:

$$\delta(x) = \begin{cases} 1, & \text{if } x = 0 \\ 0, & \text{otherwise} \end{cases} \qquad (3.4)$$

Let us now turn our attention to solving the estimation problem, $i.e.$, to estimate the state X_t given the observation history $Z_{1:t} \equiv \{Z_1, \ldots, Z_t\}$. This posterior density is given by

$$p(X_t | Z_{1:t}) = \frac{p(Z_t | X_t) p(X_t | Z_{1:t-1})}{\int p(Z_t | X_t) p(X_t | Z_{1:t-1}) dX_t}, \qquad (3.5)$$

where $p(X_t | Z_{1:t-1})$ is given by

$$p(X_t | Z_{1:t-1}) = \int p(X_t | X_{t-1}) p(X_{t-1} | Z_{1:t-1}) dX_{t-1}. \qquad (3.6)$$

Equations (3.5) and (3.6) show that knowing the posterior density $p(X_{t-1} | Z_{1:t-1})$ at $t - 1$ helps us to compute the posterior density $p(X_t | Z_{1:t})$ at t in two successive steps. Remember that the posterior density might tell us, for example, the likely position of a biological target from past (and current) observations about the target computed from the image sequence. By convention, we assume that at the beginning time $t = 0$, the posterior density is given by $p(X_0)$. Equations (3.5) and (3.6) conjunctively form a recursion beginning at $t = 0$ to compute the current posterior density. The recursion relation is known as a "sequential" formulation. The next question the reader may ask is, *why is this formulation called Bayesian?* Looking at (3.5) we may argue that as in Reverend Bayes' rule of probability, the product of a likelihood density $p(Z_t | X_t)$ and a prior density $p(X_t | Z_{1:t-1})$ and a suitable normalization factor (the denominator of (3.5)) form the posterior density $p(X_t | Z_{1:t})$. Figure 3.1 illustrates the tth recursive step of sequential Bayesian

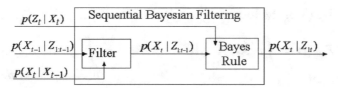

FIGURE 3.1: Recursive computation in sequential Bayesian filtering. Each stage (t) has two computation units—the filter stage and the Bayesian stage. These units are sequential. The filter unit takes in the motion model density and the posterior density of previous stage ($t - 1$) and outputs predicted posterior density $p(X_t | Z_{1:t-1})$. Next the Bayesian unit takes in this predicted posterior density and likelihood density to compute estimated posterior density $p(X_t | Z_{1:t})$ for the current stage.

filtering, along with inputs and outputs. Equations (3.6) and (3.5) are labeled as "Filter" and "Bayes Rule" respectively in Fig. 3.1.

The density $p(X_t | Z_{1:t})$ can be manipulated to form

$$p(X_t | Z_{1:t}) = p(X_t | Z_t, Z_{1:t-1}) = \frac{p(X_t, Z_t | Z_{1:t-1})}{p(Z_t | Z_{1:t-1})} = \frac{p(X_t, Z_t | Z_{1:t-1})}{\int p(X_t, Z_t | Z_{1:t-1}) dX_t}$$

$$= \frac{p(Z_t | X_t, Z_{1:t-1}) p(X_t | Z_{1:t-1})}{\int p(Z_t | X_t, Z_{1:t-1}) p(X_t | Z_{1:t-1}) dX_t}. \tag{3.7}$$

The first step in (3.7) follows from the definition of $Z_{1:t}$; the second step follows by utilizing the definition of conditional probability density (*i.e.*, $p(A|B) = \frac{p(A,B)}{p(B)}$), and the third step makes use of the definition of the marginal probability density (*i.e.*, $p(Y) = \int p(Y, X) dX$). Finally, the final step in (3.7) is made by way of the conditional probability density definition. Let us now turn our attention to the measurement probability density $p(Z_t | X_t, Z_{1:t-1})$. As already mentioned, the probability density $p(Z_t | X_t)$ for the measurement (Z_t) is conditionally independent given the current state (X_t). Thus we can write

$$p(Z_t | X_t, Z_{1:t-1}) = p(Z_t | X_t). \tag{3.8}$$

Substituting $p(Z_t | X_t)$ for $p(Z_t | X_t, Z_{1:t-1})$ in (3.7) yields Eq. (3.5).

To derive (3.6) the following manipulations of $p(X_t|Z_{1:t-1})$ are performed:

$$p(X_t|Z_{1:t-1}) = \int p(X_t, X_{t-1}|Z_{1:t-1})dX_{t-1} = \int p(X_t|X_{t-1}, Z_{1:t-1})$$
$$p(X_{t-1}|Z_{1:t-1})dX_{t-1}. \quad (3.9)$$

The first step in (3.9) uses the definition of marginal probability density and the second step follows from the definition of conditional probability density as before. Now we need to apply the Markovian assumption on the state transition density $p(X_t|X_{t-1})$, *i.e.* the probability of the current state (X_t) is conditionally independent given the previous state (X_{t-1}):

$$p(X_t|X_{t-1}, Z_{1:t-1}) = p(X_t|X_{t-1}). \quad (3.10)$$

Substituting $p(X_t|X_{t-1})$ for $p(X_t|X_{t-1}, Z_{1:t-1})$ in (3.9) results in (3.6).

Simultaneously with the recursion given jointly by (3.5) and (3.6), we can estimate the current target state via a *maximum a posteriori* (MAP) approach:

$$\hat{X}_t = \max_{X_t} p(X_t|Z_{1:t}). \quad (3.11)$$

We now have a very general framework for tracking biomedical objects from a probabilistic point of view. In the next section we discuss Kalman filter—a special case of the aforementioned sequential Bayesian estimation problem.

3.3   KALMAN FILTER

The two equations (3.5) and (3.6) form the foundation of the sequential Bayesian tracking framework. The concern is, of course, how to actually perform the recursive computation of the posterior density by Eqs. (3.5) and (3.6). There are other technical issues as well, such as whether the posterior density (3.5) has a closed form solution, and how fast the recursive computations can be performed. Let us first mention that the closed form solution for the posterior density $p(X_t|Z_{1:t})$ can be obtained if we restrict all the densities $p(X_0)$, $p(Z_t|X_t)$, and $p(X_t|X_{t-1})$ to be Gaussian. When the densities are indeed Gaussian, we obtain a familiar friend, the Kalman filter, which is a special sequential Bayesian filter where the posterior density $p(X_t | Z_{1:t})$ also becomes Gaussian (see [15]). The Kalman filter assumes linear state and measurement evolution relationships:

$$X_t = F_t X_{t-1} + V_{t-1}, \tag{3.12}$$

and

$$Z_t = H_t X_t + N_t, \tag{3.13}$$

where F_t, H_t are matrices defining a linear mapping, V_{t-1}, and N_t are independent zero mean noise processes with covariance matrices Q_{t-1} and R_t, respectively. The Kalman filter further assumes the density $p(X_0)$, which is Gaussian. Thus, given that $p(X_0)$ is Gaussian, and given the linear relationships (3.12) and (3.13), it can be shown that the densities $p(Z_t | X_t)$ and $p(X_t | X_{t-1})$ are also Gaussian. It can be shown that (3.6) and (3.5) become, respectively

$$p(X_t | Z_{1:t-1}) = G(X_t; \mu_{t|t-1}, P_{t|t-1}), \tag{3.14}$$

and

$$p(X_t | Z_{1:t}) = G(X_t; \mu_{t|t}, P_{t|t}) \tag{3.15}$$

where $G(X; \mu, P)$ is a multivariate Gaussian density with mean vector μ and covariance matrix P:

$$G(X; \mu, P) = \frac{\exp[-\frac{1}{2}(X - \mu)^T P^{-1}(X - \mu)]}{\sqrt{(2\pi)^n |P|}}, \tag{3.16}$$

where $|P|$ is the determinant of the covariance matrix P and n is the length of the vector X. The mean vectors and the covariance matrix actually follow recursions. The mean and the covariance of the time-updated density, $p(X_t | Z_{1:t-1})$, are respectively

$$\mu_{t|t-1} = F_t \mu_{t-1|t-1}, \tag{3.17}$$

and

$$P_{t|t-1} = Q_{t-1} + F_t P_{t-1|t-1} F_t^T. \tag{3.18}$$

The Gaussian measurement updated density or the posterior density $p(X_t | Z_{1:t})$ has the mean and standard deviation as

$$\mu_{t|t} = \mu_{t|t-1} + K_t(Z_t - H_t \mu_{t|t-1}), \qquad (3.19)$$

and

$$P_{t|t} = P_{t|t-1} - K_t H_t P_{t|t-1}, \qquad (3.20)$$

where

$$K_t = P_{t|t-1} H_t^T (H_t P_{t|t-1} H_t^T + R_t)^{-1} \qquad (3.21)$$

is the Kalman gain matrix. Note that when we perform the MAP estimation on the posterior Gaussian density $p(X_t | Z_{1:t})$, the estimated state we obtain is the mean vector $\mu_{t|t}$. The recursive updating starts at $t = 1$ with known (assumed) $\mu_{0|0}$ and $P_{0|0}$ (*i.e.*, with density $p(X_0) = N(X_0; \mu_{0|0}, P_{0|0})$). In addition to being the minimum mean squared estimator for this tracking scenario, the Kalman filter also holds bragging rights in computational expense: only a few matrix and vector multiplications and inversions are required at each time.

Figure 3.2 shows computations in the Kalman filter. One might compare Fig. 3.2 with the general sequential Bayesian filter computations shown in Fig. 3.1.

3.4 CASE STUDY: THE ALPHA–BETA FILTER

For the reader who is not an expert in control theory, linear algebra, and stochastic processes, the typical question at this point is *How the heck do I use the Kalman filter in an actual tracking application?* The typical text leaves the hungry reader unsatisfied at this point. Perhaps, the authors hypothesize, the reason is simple job security. *If any bubba could implement the Kalman filter, we'd lose all those lucrative consulting deals . . . .*

First, perhaps we need more intuition with respect to the Kalman filter operation. Hopefully, this section on the alpha–beta implementation will provide this intuition.

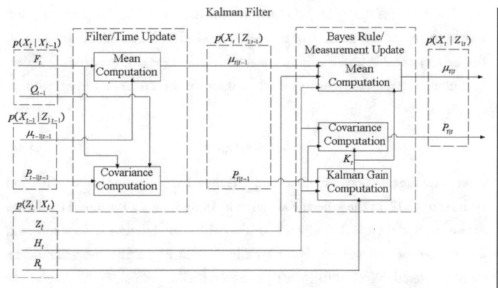

FIGURE 3.2: The Kalman Filter as a sequential Bayesian filter. The working principle is same as the sequential Bayesian Filtering. All the densities are Gaussian here; so they can be represented by their means and covariances. Consequently each of the time and measurement update units computes mean and covariance. In the measurement update unit we also have Kalman gain computation. Kalman gain is required in computing both mean and covariance in the measurement update. The dotted boxes represent a Gaussian density specified by mean and the covariance as shown.

We need to recognize that the Kalman filter is a combination of a predictor and a filter. The predictor estimates the location of the target at time t given $t - 1$ observations. When observation t arrives, the estimate is improved using an optimal filter to estimate the target position at time $t + 1$: the filtered estimate is essentially the best estimate of the true location of the target given t observations at time t. (Note: for convenience in this section, we use the notation of time $t + 1$ for the prediction, and time t for the filtered estimate.)

In this section, we develop a tracker that assumes a constant velocity model for targets. In many biomedical imaging applications, this assumption is reasonable, at least within small segments of time. This tracker is called the alpha–beta filter, as the two Kalman gains needed are α and β, which we will run into shortly. In this description of the alpha–beta filter, we will develop a filter for prediction and estimation of the row position of the target, denoted by i. The column position, j, can be computed in an identical manner with an independently operating alpha–beta filter. The constant velocity model for the alpha–beta filter for vertical position i is given by

$$i_{t+1} = i_t + \delta t v_t^i \tag{3.22}$$

where δt is the time between frames and v_t^i is the velocity at time t in the i-direction (similar to (3.12) in the general Kalman model). By applying the model (3.22), we are saying that the first derivative of position with respect to time is fairly constant, and the second derivative is almost zero. In the alpha–beta filter, we will model a small acceleration (called the "drift") as a white noise process. So, as long as our temporal sampling rate is sufficient, this should be a good model for motion in most Newtonian circumstances. Note that our approach is to model acceleration as a noise process with a reasonably chosen variance.

We could, instead, use a constant acceleration model with the Kalman filter—this is the *alpha–beta–gamma* filter. This model has the acceleration terms in addition to position and velocity (for each direction, i and j). Keep in mind that unless one is certain of the model, adding extra (dubious) terms to the Kalman model makes prediction worse in general.

Now, back to the alpha–beta filter implementation—the prediction for the vertical position i is then (similar to the general model of (3.17))

$$\hat{i}_{t+1|t} = \hat{i}_{t|t} + \delta t \, \hat{v}_{t+1|t}^i \tag{3.23}$$

where $\hat{v}_{t+1|t}^i$ is the predicted velocity (to be derived below) and $\hat{i}_{t|t}$ is a (filtered) estimate of the observed i_t at time t:

$$\hat{i}_{t|t} = \hat{i}_{t|t-1} + \alpha_t \big(i_t^0 - \hat{i}_{t|t-1} \big) \tag{3.24}$$

where α_t is a gain. Note that observed horizontal position i_t^0 is corrupted by observation noise: $i_t^0 = i_t + n_t$, where n_t is a random variable representing the unknown observation noise (in the i-direction) at time t. Note that (3.24) is the alpha–beta version of (3.19). Intuition is gained by rearranging (3.24) as $\hat{i}_{t|t} = (1 - \alpha_t)\hat{i}_{t|t-1} + \alpha_t i_t^0$. Notice that the gain α_t determines the balance between the previous track history and the new observation from the image when we calculate $\hat{i}_{t|t}$. If α_t is large (near 1), we believe the observations are very reliable (essentially ignoring the track history). If α_t is small (near 0), we believe that there is significant corruption from measurement noise (essentially ignoring the observation). In other words, the larger the variance of the measurement noise, the more we want to depend on the history, connoting lower values for α_t.

In this alpha–beta implementation, the velocity is modeled as

$$v_{t+1}^i = v_t^i + u_t^i, \tag{3.25}$$

which implies that velocity is constant except for a random drift u_t^i drawn from a Gaussian distribution. Then, the prediction for the velocity at time $t + 1$ is

$$\hat{v}_{t+1|t}^i = \hat{v}_{t|t-1}^i + \beta_t\big(i_t^0 - \hat{i}_{t|t-1}\big)/\delta t, \tag{3.26}$$

where β_t is the second gain needed for the alpha–beta filter implementation. β_t controls how we let the new observation affect the predicted velocity. If β_t is near 0, it means that we think the observations are unreliable and that the actual velocity is really constant—in this case, the observation does not affect the predicted velocity. If β_t approaches unity, then the observations are reliable (we think!). In this case, we are allowing the velocity to drift (which is in effect enacting acceleration).

3.4.1 Alpha–Beta Filter Gains

Assuming a stationary model, we can pre-compute the gains in the alpha–beta filter even before the first image is acquired. Also, since the gains converge quickly to constants, we do not have to compute an infinite number of the gain values. Then, implementation of the alpha–beta tracker becomes simply a matter of plugging the observations into the correct equations, which can yield real-time predications and estimates of position and velocity.

Recall from the general discussion on the Kalman filter that the Kalman gains depend upon the **noise variances** and the **state vector error covariance matrix**. Let the state vector X be defined for our system as $X_t = \begin{bmatrix} i_t \\ v_t^i \end{bmatrix}$. The state vector of the predictor is then $\mu_{t|t-1} = \begin{bmatrix} \hat{i}_{t|t-1} \\ \hat{v}_{t|t-1}^i \end{bmatrix}$, and the state vector for the filter $\mu_{t|t}$ is constructed similarly. Of course, these are the terms needed for the alpha–beta filter for the vertical position and velocity; once again, we mention that corresponding terms must be constructed for the horizontal direction (and possibly in a third direction if tracking in 3D).

The overall goal is to provide a prediction of the state at time $t+1$ and a refined estimate of the state at time t with minimal error. The error in the predicted state vector is $X_t - \mu_{t|t-1}$, and the error for the filter state vector is $X_t - \mu_{t|t}$.

These error terms are stochastic vectors, hence they have covariance matrices. The predicted state vector covariance matrix is $P_{t|t-1} = E[(X_t - \mu_{t|t-1})(X_t - \mu_{t|t-1})^T]$. The filtered state vector error covariance matrix is $P_{t|t} = E[(X_t - \mu_{t|t})(X_t - \mu_{t|t})^T]$. Essentially, the **function of the Kalman filter** in this alpha–beta implementation is to choose α_t and β_t such that $P_{t|t}$ is minimized.

The solution that in fact minimizes $P_{t|t}$ is [16]

$$\alpha_t = \frac{P_{t|t-1}^{11}}{P_{t|t-1}^{11} + \sigma_n^2}, \tag{3.27}$$

where σ_n^2 is the measurement noise variance, $P_{t|t-1}^{11}$ is the upper-left element of the 2×2 matrix $P_{t|t-1}$, and

$$\beta_t = \frac{\delta t P_{t|t-1}^{21}}{P_{t|t-1}^{11} + \sigma_n^2}. \tag{3.28}$$

At this point, we have equations for α_t and β_t; however, $P_{t|t-1}$ is unknown. For our constant velocity alpha–beta model, $P_{t|t-1}$ can be computed recursively as follows:

$$P_{t+1|t}^{11} = P_{t|t-1}^{11} + 2P_{t|t-1}^{12} + P_{t|t-1}^{22} - \frac{\left(P_{t|t-1}^{11} + P_{t|t-1}^{12}\right)^2}{P_{t|t-1}^{11} + \sigma_n^2},$$

$$P_{t+1|t}^{12} = P_{t|t-1}^{12} + P_{t|t-1}^{22} - P_{t|t-1}^{12}\left(\frac{P_{t|t-1}^{11} + P_{t|t-1}^{12}}{P_{t|t-1}^{11} + \sigma_n^2}\right),$$

$$P_{t+1|t}^{21} = P_{t+1|t}^{12}$$

and

$$P_{t+1|t}^{22} = P_{t|t-1}^{22} + \sigma_u^2 - \frac{\left(P_{t|t-1}^{12}\right)^2}{P_{t|t-1}^{11} + \sigma_n^2}. \tag{3.29}$$

We now have the basic tools in hand to track a moving object in a given biomedical application using the alpha–beta implementation of the Kalman filter. Nevertheless, we must first discuss initialization.

3.4.2 Initializing the Kalman Tracker

First, we need initial conditions for $P_{t|t-1}$ (*viz.*, $P_{1|0}$). This particular subject has caused controversy on many a white board, so we ask the reader to keep in mind that there is no universally accepted approach. To make this controversial initialization, we need two additional variances: σ_i^2, the variance in the initial row position (there is a corresponding term for column position j), and $\sigma_{v_i}^2$, the variance in the initial velocity in the i-direction.

We could assume, for example, that the initial position is a uniformly distributed random variable over the possible rows. The computation of $\sigma_{v_i}^2$ could be carried out in the same way—by determining the minimum and maximum possible velocities and then assuming that velocity is uniformly distributed over the possible velocities.

Given these two additional terms, we can compute the filtered state vector error covariance at time $t = 0$:

$$P_{0|0} = E[(X_0)(X_0)^T] = \begin{bmatrix} \sigma_i^2 & 0 \\ 0 & \sigma_{v_i}^2 \end{bmatrix}, \tag{3.30}$$

and $P_{1|0}$ can be computed from

$$P_{1|0} = F_0 P_{0|0} F_0^T + Q_0, \tag{3.31}$$

where $F_0 = \begin{bmatrix} 1 & \delta t \\ 0 & 1 \end{bmatrix}$ is the state transition matrix, and $Q_0 = \begin{bmatrix} \sigma_n^2 & 0 \\ 0 & \sigma_u^2 \end{bmatrix}$ gives the initial covariance of the two noise processes.

Typically, the initial filtered position, $\hat{i}_{0|0}$, is set to the first observed position, and the first velocity estimate is left indeterminate or is set to an arbitrary constant.

3.4.3 Executing the Alpha–Beta Filter

In summary, the alpha–beta filter implementation requires the following steps:

1. Use the starting state conditions to get the initial $P_{t|t-1}$ matrix.

2. Use (3.27) and (3.28) to acquire alpha and beta for several ts until convergence is achieved, and store these values.

3. Acquire the target and obtain initial coordinates. (For example, use thresholding or a centroid calculation or an active contour to find the target. We provide two such methods in the next subsection.)

4. Use (3.24) to compute the filtered position $\hat{i}_{t|t}$. Then, use (3.26) to compute the predicted velocity $\hat{v}_{t+1|t}^i$, and finally apply (3.23) to get the predicted position $\hat{i}_{t+1|t}$ (and repeat for the corresponding terms for the j-direction).

5. Acquire target within track gate (a subimage) centered at predicted position.

6. Go to step 4.

If there are no observations at time t, the track is *coasted*. In such a case, we use $i_t^0 = \hat{i}_{t|t-1}$; the observation is replaced by the last prediction.

3.4.4 Obtaining Measurements for the Alpha–Beta Filter

3.4.4.1 Template Matching

There are several viable and time-tested methods for obtaining observations for the alpha–beta filter (and for other trackers as well). Here, we consider two simple solutions—the correlation method and the centroid method. In the correlation approach, an old friend of cold warriors and beltway bandits, we first define an object template $T(m, n)$ of height $(2h + 1)$ and width $(2w + 1)$. This template could be, for example, a subimage containing the manually identified target, taken from the first frame of a video sequence. At time t, after making the prediction, a window (a rectangular region in the image) $M(\hat{i}_{t|t-1}, \hat{j}_{t|t-1})$ from the tth video frame centered at the prediction vector $(\hat{i}_{t|t-1}, \hat{j}_{t|t-1})$ is considered. Next the measurement vector is computed via maximizing normalized cross-correlation between the target template and the video frame:

$$
(i_t^0, j_t^0) = \underset{k,l}{\arg\max} \left(\frac{\displaystyle\sum_{m=k-h,n=l-w}^{m=k+h,n=l+w} [I_t(m, n) - \bar{I}_t(k, l)][T(m-k, n-l) - \overline{T}]}{\sqrt{\displaystyle\sum_{m=k-h,n=l-w}^{m=k+h,n=l+w} [I_t(m, n) - \bar{I}_t(k, l)]^2 \ \sum_{i=-h,j=-w}^{i=h,j=w} [T(m, n) - \overline{T}]^2}} \right),
$$

(3.32)

where $\bar{I}_t(k, l)$ is the mean image intensity defined as

$$
\bar{I}_t(k, l) = \frac{1}{(2h + 1)(2w + 1)} \sum_{m=k-h,n=l-w}^{m=k+h,n=l+w} I_t(m, n),
$$

(3.33)

and $\overline{T}$ is mean template intensity:

$$
\bar{T}(k, l) = \frac{1}{(2h + 1)(2w + 1)} \sum_{m=-h,n=-w}^{m=h,n=w} T(m, n).
$$

(3.34)

While finding the argument that produces a maximal value in (3.32), only those values of (k, l) belonging to the window $M(\hat{i}_{t|t-1}, \hat{j}_{t|t-1})$ are taken into account.

In addition to correlation, template matching can be achieved using less expensive measures. Many algorithms have subsisted by measuring the sum of absolute differences (between the template and the subimage at hand) or the sum of squared differences. It is only the normalized cross-correlation measure, however, that guarantees a minimum mean squared error solution.

3.4.4.2 Centroid Measurements

Instead of a correlation measurement, one may opt for a centroid measurement. Centroid tracking became popular in the sixties and seventies for tracking bright targets on dark backgrounds, such as missile plumes in the sky, or tank engines from infrared images. The centroid tracker can also have biomedical application (e.g., bright cells on a dark background). Centered at the predicted position, $(\hat{i}_{t|t-1}, \hat{j}_{t|t-1})$, a window $M(\hat{i}_{t|t-1}, \hat{j}_{t|t-1})$ of height $(2H+1)$ and width $(2W+1)$ is extracted. Then the centroid of the image intensity within this window is computed and the centroid is taken as the measurement vector:

$$i_t^0 = \hat{i}_{t|t-1} + \frac{\sum\limits_{m=-H,n=-W}^{m=H,n=W} m I_t(m, n)}{\sum\limits_{m=-H,n=-W}^{m=H,n=W} I_t(m, n)} \tag{3.35}$$

and

$$j_t^0 = \hat{j}_{t|t-1} \frac{\sum\limits_{m=-H,n=-W}^{m=H,n=W} n I_t(m, n)}{\sum\limits_{m=-H,n=-W}^{m=H,n=W} I_t(m, n)} \tag{3.36}$$

These calculations give the centroid for a "hot" target. For a "cold" target (a dark target on a bright background), one can replace the intensity $I_t(m, n)$ with $I_{max} - I_t(m, n)$, where I_{max} is the maximum possible intensity.

The size of window M can be determined through an analytical way, known as *gating*. Note that the prediction density in the Kalman filter is Gaussian and is as follows:

$$p((i_t, j_t)|Z_{1:t-1}) = \frac{\exp\left(-\frac{1}{2}\left(\begin{bmatrix} i_t \\ j_t \end{bmatrix} - \mu_{t|t-1}\right)^T P_{t|t-1}^{-1}\left(\begin{bmatrix} i_t \\ j_t \end{bmatrix} - \mu_{t|t-1}\right)\right)}{2\pi\sqrt{|P_{t|t-1}|}}. \tag{3.37}$$

The random variable $\left(\left[\begin{smallmatrix} i_t \\ j_t \end{smallmatrix}\right] - \mu_{t|t-1}\right)^T P_{t|t-1}^{-1} \left(\left[\begin{smallmatrix} i_t \\ j_t \end{smallmatrix}\right] - \mu_{t|t-1}\right)$ is Chi-squared distributed with two degrees of freedom [17]. So, the window area can be obtained as

$$\left(\begin{bmatrix} x_t \\ y_t \end{bmatrix} - \mu_{t|t-1}\right)^T P_{t|t-1}^{-1} \left(\begin{bmatrix} x_t \\ y_t \end{bmatrix} - \mu_{t|t-1}\right) \leq \gamma \tag{3.38}$$

where in this 2D case, one can choose the gate size parameter γ by choosing a desired probability value $1 - \exp(-\gamma/2)$, which reflects our belief that the measurement will be found within the elliptical gate with probability $1 - \exp(-\gamma/2)$.

Figure 3.3 shows tracking leukocytes from intravital video microscopy with centroid measurements and the alpha–beta Kalman filter.

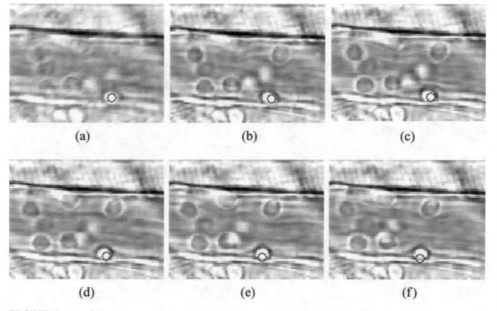

FIGURE 3.3: Tracking a cell with the Kalman filter (alpha–beta filter) with the centroid as the measurement (detection) mechanism. (a) and (b) Six consecutive frames. Tracked cell center indicated by circles.

3.5   THE EXTENDED KALMAN FILTER

One of the limiting assumptions connected with the Kalman filter approach is linearity. In cases where the model is moderately nonlinear, the Extended Kalman filter (EKF) may be applied. Here, the motion model becomes

$$X_t = f_{t-1}(X_{t-1}, V_{t-1}), \tag{3.39}$$

and the observation model is written as

$$Z_t = h_t(X_t, N_t), \tag{3.40}$$

where f and h are the nonlinear functions. The terms X, Z, V, and N carry the identical meaning as in the Kalman filter case—X is the state, Z is the observation, V is the process noise, and N is the measurement noise. As with the Kalman filter, we want to estimate X_t at the current time point t given all the measurements $Z_{1:t}$ available at this time.

The EKF approximates (linearizes) (3.39) and (3.40) via Taylor series expansion (3.39) becomes

$$X_t \approx f_t(\mu_{t-1|t-1}, 0) + F_t(X_{t-1} - \mu_{t-1|t-1}) + A_t V_{t-1}, \tag{3.41}$$

and (3.40) becomes

$$Z_t \approx h_t(\mu_{t|t-1}, 0) + H_t(X_t - \mu_{t|t-1}) + B_t N_t. \tag{3.42}$$

F, A, H, and B are Jacobian matrices and are defined as follows:

$$F_t(m, n) = \frac{\partial f_t(m)}{\partial X_{t-1}(n)}(\mu_{t-1|t-1}, 0), \tag{3.43}$$

$$A_t(m, n) = \frac{\partial f_t(m)}{\partial V_{t-1}(n)}(\mu_{t-1|t-1}, 0), \tag{3.44}$$

$$H_t(m, n) = \frac{\partial h_t(m)}{\partial X_t(n)}(\mu_{t|t-1}, 0), \tag{3.45}$$

and

$$B_t(m, n) = \frac{\partial h_t(m)}{\partial N_t(n)}(\mu_{t|t-1}, 0).$$ (3.46)

For example, $A_t(m, n)$ denotes the (m, n)th element of the Jacobian matrix, $h_t(m)$ denotes the mth element of the vector nonlinear function, and so on. The approximations (3.41) and (3.42) of course depend on the degree of nonlinearity involved in the functions f and h. To make this point clear, let us give an example. If f is a second order polynomial then the approximation (3.41) is better than if it were a cubic polynomial. Accordingly, the lower the degree of nonlinearity, the better the EKF performance.

After achieving the linear relationships (3.41) and (3.42) via first order Taylor series approximation, all the theory behind Kalman filter can now be applied to this estimation problem. Once again, the prediction density and the posterior density are Gaussian as before. The prediction density mean and covariance are given respectively by

$$\mu_{t|t-1} = f_{t-1}(\mu_{t-1|t-1}, 0),$$ (3.47)

and

$$P_{t|t-1} = F_t Q_{t-1} F_t^T + A_t Q A_t^T.$$ (3.48)

Similarly the posterior density is determined by the mean vector

$$\mu_{t|t} = \mu_{t|t-1} + K_t(Z_t - h_t(\mu_{t|t-1}, 0)),$$ (3.49)

and the covariance matrix

$$P_{t|t} = P_{t|t-1} - K_t H_t P_{t|t-1},$$ (3.50)

where

$$K_t = P_{t|t-1} H_t^T \left(H_t P_{t|t-1} H_t^T + B_t R_t B_t^T\right)^{-1}$$ (3.51)

is the Kalman gain matrix. Notice that the state estimation recursion equations are slightly different in form than with the original Kalman filter.

EKF provides a clever solution when the motion model and the measurement model are nonlinear. Often, complex interactions and non-Newtonian behavior in biological systems leads to such nonlinearity. On the other hand, if the noise processes are non-Gaussian or if the degree of nonlinearity is significant, then particle filters come to the rescue (see Chapter 4).

3.6   INTERACTING MULTIPLE MODELS FOR TRACKING

The assumptions regarding the motion model in a tracking problem are critical. However we often notice in applications that it is difficult to define object motions by a single probability distribution. Typical examples are intermittent motion, combination of smooth motion and sudden jump, and the like. In these situations a single probability distribution describing the motion of an object is insufficient. Instead, multiple motion models to describe the motion of such an object are quite helpful. Thus the aim of this section is to describe how we can probabilistically formulate tracking with multiple motion models.

In the spirit of the interacting multiple model (IMM) introduced by Blom and Bar-Shalom [18], we assume that M motion models are available and that we can represent the choice of the model with a Markov chain with state θ_t at time t. So, when $\theta_t = 3$, the third motion model (of M choices) is in effect at time t. We have in hand a state transition probability of $p(X_t|X_{t-1}, \theta_t)$, $t \geq 1$, and a measurement probability of $p(Z_t|X_t, \theta_t)$, $t \geq 1$. Note that these probabilities are assumed to be model dependent. Another assumption is that the M-by-M state transition matrix expressing Markov transition probabilities, $p(\theta_t|\theta_{t-1})$, $t \geq 1$, is also known.

We now turn our attention to the state estimation problem as in Section 3.1. Remember that our overall goal is to compute the posterior probability density $p(X_t|Z_{1:t})$ in order to estimate the target state X_t. Now let us gradually unravel the construction of the posterior density. We can write this density (using the rule of marginal probability densities) as

$$p(X_t | Z_{1:t}) = \sum_{\theta_t = 1}^{M} p(X_t, \theta_t | Z_{1:t}) = \sum_{\theta_t = 1}^{M} p(X_t | \theta_t, Z_{1:t}) p(\theta_t | Z_{1:t}). \qquad (3.52)$$

To compute the posterior $p(X_t | Z_{1:t})$ from (3.52), we need to compute two densities $p(X_t | \theta_t, Z_{1:t})$ and $p(\theta_t | Z_{1:t})$.

In Fig. 3.4, we illustrate one cycle of the recursive IMM filtering for time t. In the figure, the probabilities and the densities computed within the previous time $(t - 1)$ are shown within hexagonal boxes. The outputs are denoted within oval boxes. Note that these outputs become the inputs for the next time $(t + 1)$. We assume the availability of the state transition probability and the measurement probability at each time. As the figure illustrates, the entire IMM recursive

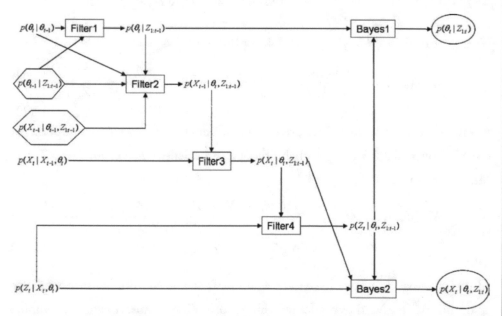

FIGURE 3.4: IMM filtering has six computational units—*Filter*1 through *Filter*4 and *Bayes*1 and *Bayes*2. The corresponding input and output densities are shown by arrows to and from each unit, respectively. Densities shown in hexagonal boxes are outputs from the previous stage $(t - 1)$, whereas densities shown in the ovals are outputs from this stage (t).

estimation comprises of four filtering steps and two steps implementing Bayes' rule. Let us now look at each of these six steps more closely.

*Filter*1 is derived from the total probability rule as follows:

$$P(\theta_t | Z_{1:t-1}) = \sum_{\theta_{t-1}=1}^{M} P(\theta_t | \theta_{t-1}, Z_{1:t-1}) P(\theta_{t-1} | Z_{1:t-1})$$

$$= \sum_{\theta_{t-1}=1}^{M} P(\theta_t | \theta_{t-1}) P(\theta_{t-1} | Z_{1:t-1}), \tag{3.53}$$

where in the second equality we utilize the conditional independence assumption of density (3.51).

To derive *Filter*2, we once again start with the rule of total probability applied to $p(X_{t-1} | \theta_t, Z_{1:t-1})$ as

$$p(X_{t-1} | \theta_t, Z_{1:t-1}) = \sum_{\theta_t=1}^{M} p(X_{t-1} | \theta_{t-1}, \theta_t, Z_{1:t-1}) p(\theta_{t-1} | \theta_t, Z_{1:t-1}). \tag{3.54}$$

Next we can write $p(\theta_{t-1} | \theta_t, Z_{1:t-1})$ using Bayes' rule as

$$p(\theta_{t-1} | \theta_t, Z_{1:t-1}) = \frac{p(\theta_t | \theta_{t-1}, Z_{1:t-1}) p(\theta_{t-1} | Z_{1:t-1})}{p(\theta_t | Z_{1:t-1})} = \frac{p(\theta_t | \theta_{t-1}) p(\theta_{t-1} | Z_{1:t-1})}{p(\theta_t | Z_{1:t-1})} \tag{3.55}$$

where in the second equality we used the conditional independence assumption for the Markov model switching probabilities. Now the density $p(X_{t-1} | \theta_{t-1}, \theta_t, Z_{1:t-1})$ can be written as

$$p(X_{t-1} | \theta_{t-1}, \theta_t, Z_{1:t-1}) = \frac{p(\theta_t | \theta_{t-1}, X_{t-1}, Z_{1:t-1}) p(X_{t-1} | \theta_{t-1}, Z_{1:t-1})}{p(\theta_t | \theta_{t-1}, Z_{1:t-1})}$$

$$= p(X_{t-1} | \theta_{t-1}, Z_{1:t-1}). \tag{3.56}$$

In establishing (3.56), we have once again utilized the conditional independence assumption of the Markov model switching probabilities. Thus we can now combine (3.55) and (3.56) with (3.54) to obtain an expression for *Filter*2:

$$p(X_{t-1} | \theta_t, Z_{1:t-1}) = \sum_{\theta_t=1}^{M} \frac{p(X_{t-1} | \theta_{t-1}, Z_{1:t-1}) p(\theta_t | \theta_{t-1}) p(\theta_{t-1} | Z_{1:t-1})}{p(\theta_t | Z_{1:t-1})}. \tag{3.57}$$

*Filter*3 is derived by utilizing the rule of marginal probability density applied to $p(X_t|\theta_t, Z_{1:t-1})$ as

$$p(X_t|\theta_t, Z_{1:t-1}) = \int p(X_t, X_{t-1}|\theta_t, Z_{1:t-1})dX_{t-1}$$
$$= \int p(X_t|X_{t-1}, \theta_t, Z_{1:t-1})p(X_{t-1}|\theta_t, Z_{1:t-1})dX_{t-1}.$$
$$= \int p(X_t|X_{t-1}, \theta_t)p(X_{t-1}|\theta_t, Z_{1:t-1})dX_{t-1} \qquad (3.58)$$

In the equality of (3.58), we have utilized the conditional independence of the state transition density.

To derive *Filter*4 we once use the rule for marginal density as follows, followed by Bayes' rule to obtain

$$P(Z_t|\theta_t, Z_{1:t-1}) = \int P(Z_t, X_t|\theta_t, Z_{1:t-1})dX_t$$
$$= \int p(Z_t|X_t, \theta_t, Z_{1:t-1})p(X_t|\theta_t, Z_{1:t-1})dX_t,$$
$$= \int p(Z_t|X_t, \theta_t)p(X_t|\theta_t, Z_{1:t-1})dX_t \qquad (3.59)$$

where in obtaining the last equality we use the conditional independence of the measurement (or the likelihood) density (note that we assumed a similar conditional independence in Section 3.2).

Finally we are left to spell out *Bayes*1 and *Bayes*2. *Bayes*1 is nothing other than

$$p(\theta_t|Z_{1:t}) = \frac{P(Z_t|\theta_t, Z_{1:t-1})P(\theta_t|Z_{1:t-1})}{\sum\limits_{\theta_t=1}^{M} P(Z_t|\theta_t, Z_{1:t-1})P(\theta_t|Z_{1:t-1})}, \qquad (3.60)$$

and similarly *Bayes*2 implements

$$p(X_t|\theta_t, Z_{1:t}) = \frac{p(Z_t|X_t, \theta_t, Z_{1:t-1})p(X_t|\theta_t, Z_{1:t-1})}{p(Z_t|\theta_t, Z_{1:t-1})}$$
$$= \frac{p(Z_t|X_t, \theta_t)p(X_t|\theta_t, Z_{1:t-1})}{p(Z_t|\theta_t, Z_{1:t-1})}, \qquad (3.61)$$

where the first equality is the result of applications of the definition of conditional probability, and in deriving the second equality, we utilize the conditional independence assumption associated with the measurement density.

Let us now quickly look at the assumptions we have made in deriving the IMM recursions. We only have used three conditional independence assumptions associated with the state transition probabilities (for the individual motion models), the measurement probabilities, and the Markov switching probabilities. For example, the conditional independence of the Markov switching probability implies that given θ_{t-1}, the probability of θ_t is independent of any other random variable. Similarly, conditional independence for the state transition mandates that given X_{t-1} and θ_t, the density of X_t is independent of any other random variable. Notice that all the filtering equations and Bayes' rule equations are derived in terms of general probability distributions—in contrast to the Kalman filter construction, where no assumptions about the densities are made.

So how is IMM tracking initiated? By assumption, even at time $t = 1$, we know the measurement probabilities, the state transition probabilities, and the Markov switching probabilities. That leaves us with two unresolved inputs: (a) $p(\theta_0 | Z_{1:0})$ and (b) $p(X_0 | \theta_0, Z_{1:0})$. First let us note that in the beginning, there are no measurements available, so $p(\theta_0 | Z_{1:0})$, and $p(X_0 | \theta_0, Z_{1:0})$ are actually $p(\theta_0)$ and $p(X_0 | \theta_0)$ respectively. As with any typical Markov chain initiation, these initial distributions must be assumed for a particular application. The values of $p(\theta_0)$ are simply the probabilities of the M models, which may be assigned from extensive observations or from domain knowledge about the particular problem; often, a uniform distribution is assumed. Expressions for $p(X_0 | \theta_0)$ can also be derived from initial object detection results. As in Section 3.2, we can assume a singular distribution at (x_u, y_u) where (x_u, y_u) is the detected target position within the 0th frame.

Why is this tracking technique "interacting"? Let us take a look at Fig. 3.4. The output at each state is essentially M probability densities $p(X_t | \theta_t, Z_{1:t})$ and M probability values $p(\theta_t | Z_{1:t})$. We can think of the entire filtering as creating M

densities, $p(X_t|\theta_t, Z_{1:t})$, one for each motion model. The interaction among them actually comes through the Markov state transition matrix (see Fig. 3.4).

From this general sequential Bayesian framework for the IMM tracking mechanism, we can easily obtain results for the Kalman filter (if the state transition and measurement probabilities are both Gaussian). In the Kalman case, the state transitions and the measurement equations are linear as follows:

$$X_t = F_t(\theta_t)X_{t-1} + V_{t-1}, \tag{3.62}$$

and

$$Z_t = H_t(\theta_t)X_t + N_t, \tag{3.63}$$

with V_{t-1} and N_t being i.i.d. zero mean Gaussian noise. The matrices F_t and H_t now depend on motion model θ_t. This is the setting for the Kalman filter.

Once again we comment that the six computational steps shown in Fig. 3.4 will yield closed form solutions in special cases as with the Kalman filter. *What about other cases?* Well, for that answer, we need to learn the workings of the particle filter, which is described in the next chapter.

3.7   SUMMARY

This chapter focused on the sequential Bayesian approach to tracking. We have in our grasp the basic general principles behind sequential Bayesian tracking and the classical Kalman filter, the alpha–beta filter, and the extended Kalman filter. It is essential that the biomedical image analysis researcher master the concepts of the Bayesian solution. If linear models and Gaussian distributions for measurement and process noise are appropriate, then the Kalman filter is a priceless tool for tracking. In scenarios where these assumptions do not hold, we can still apply the approach of Reverend Bayes—but in this case, through a particle filter. The next chapter focuses on this class of sampling solutions.

CHAPTER 4

Particle Filters and Multi-Target Tracking

"In the seventies, you might have said 'I'm using the FFT,' and everyone knew what you meant. That was enough. Today, we say, 'I'm using the particle filter.' That is enough."

—Rama Chellappa

"People with no statistics background insist on rediscovering solved problems."

—Ed Delp

4.1 OVERVIEW

Despite the objections raised by Kalman in the introduction to Chapter 3, we explore sampling—essentially a weighted guessing game—to accommodate tracking scenarios where the uncertainties are not normal and where the dynamics are highly nonlinear. In the biomedical arena, we often encounter corruptive processes that are not well modeled by the distribution named for Carl Frederich Gauss and not well modeled by dynamics that are not captured by the linear motion model of Sir Isaac Newton. Here, the FFT of the new millennium, the so-called particle filter, is introduced as an indispensable tool for tracking biomedical objects.

We also attack multi-target tracking in this chapter. Fortuitously, the particle filter framework allows the incorporation of complex target-to-target interactions.

4.2   THE PARTICLE FILTER

In the Bayesian tracking framework, we would like to discover the target state X (e.g., position, velocity, shape, *etc.*) from the image observations contained in Z. Consequently, our mathematical goal is the reconstruction of the unknown posterior density $p(X_t \mid Z_{1:t})$. Often the closed form solution for the posterior density cannot be obtained; for example, when one of the densities $[p(X_0), p(X_t \mid X_{t-1})$, or $p(Z_t \mid X_t)]$ is non-Gaussian, there is no closed form solution for $p(X_t \mid Z_{1:t})$, which is often the case in biomedical tracking. The **particle filter** is a numerical method that approximates the posterior density by a set of M weighted samples of this density. Let the particle set $\{(s_{t-1}^1, w_{t-1}^1), \ldots, (s_{t-1}^M, w_{t-1}^M)\}$ represent the posterior density for time $t-1$:

$$p(X_{t-1} \mid Z_{1:t-1}) \approx \sum_{i=1}^{M} w_{t-1}^i \delta(X_{t-1} - s_{t-1}^i), \qquad (4.1)$$

where δ represents the Kroneker delta function, and the weights sum to unity: $\sum_{i=1}^{M} w_{t-1}^i = 1$.

Since the s_{t-1}^ks are independent samples from the density $p(X_{t-1} \mid Z_{1:t-1})$, the representation (4.1) in fact follows from the law of large numbers for independent samples. Also the same law dictates that the approximation (4.1) becomes more and more accurate with increasing sample size M. Thus, given a particle set to represent the immediate past posterior density as (4.1), we can now write the integral for the predicted posterior density (3.6) as a sum

$$p(X_t \mid Z_{1:t-1}) = \frac{1}{M} \sum_{i=1}^{M} w_{t-1}^i p(X_t \mid X_{t-1} = s_{t-1}^i). \qquad (4.2)$$

With the particle representation (4.2), the integral for the estimated (current) posterior now takes the following form:

$$p(X_t \mid Z_{1:t}) = c_t p(Z_t \mid X_t) \sum_{i=1}^{M} w_{t-1}^i p(X_t \mid X_{t-1} = s_{t-1}^i), \qquad (4.3)$$

where the normalization factor c_t does not depend in any way on the current state X_t. We will see that this factor c_t, a sort of mathematical crutch that enables equality in (4.3), does not actually enter into any computation associated with our particle filter tracking.

Our task is now to generate a set of M particles $\{(s_t^1, w_t^1), \ldots, (s_t^M, w_t^M)\}$ from (4.3) to represent the current posterior $p(X_t \mid Z_{1:t})$. Note that once we are able to do so, the recursion continues, *i.e.*, in the next step we will generate the particle set $\{(s_{t+1}^1, w_{t+1}^1), \ldots, (s_{t+1}^M, w_{t+1}^M)\}$ from

$$p(X_{t+1} \mid Z_{1:t+1}) = c_{t+1} p(Z_{t+1} \mid X_{t+1}) \sum_{i=1}^{M} w_t^i p(X_{t+1} \mid X_t = s_t^i). \qquad (4.4)$$

This recursion provides enormous flexibility in estimating the posterior density in a broad range of biomedical tracking applications.

We will consider two methods for generating a particle set $\{(s_t^1, w_t^1), \ldots, (s_t^M, w_t^M)\}$ from the representation (4.3) and the previous particle set $\{(s_{t-1}^1, w_{t-1}^1), \ldots, (s_{t-1}^M, w_{t-1}^M)\}$. However before we introduce these two methods, let us quickly remind the reader of the purpose of generating the particle set—to infer the current state. *So, how is the inference performed once we obtain the particle set*: $\{(s_t^1, w_t^1), \ldots, (s_t^M, w_t^M)\}$? Again, the law of large number dictates that if we want to infer the state of some function $h(X_t)$, then the following operation can be performed:

$$h(X_t) \approx \frac{1}{M} \sum_{i=1}^{M} w_t^i h(s_t^i). \qquad (4.5)$$

In particular, when we are interested in the estimation of the state X_t by the expected value of X_t, we insert $h(X_t) = X_t$ in (4.5). On the other hand, if we are interested in *maximum a posteriori* (MAP) estimation, then the estimated target state is s_t^j where $j = \arg\max_j \{w_t^1, \ldots, w_t^M\}$.

4.2.1   The CONDENSATION Algorithm

One popular method for sampling from the posterior density in (4.3) is known as CONDENSATION (Conditional Density Propagation) as proposed by Blake and Isard [19]. The method is based on *factored sampling* [20]:

Let $f(x) \propto f_1(x) f_2(x)$ be a probability density function, where $f_2(x)$ in turn is another probability density function. If it is impossible to sample directly from $f(x)$, then one can approximate the density $f(x)$ by the weighted sample set $\{[s^1, f_1(s^1)/w], \ldots, [s^M, f_1(s^M)/w]\}$, where the samples s^is are independently drawn from the density $f_2(x)$ and w is a normalizing weight, *i.e.*, $w = \sum f_1(s^i)$.

CONDENSATION considers the density (4.3) as a product of two functions: $p(X_t \mid Z_{1:t-1}) = \frac{1}{M} \sum_{i=1}^{M} w_{t-1}^i p(X_t \mid X_{t-1} = s_{t-1}^i)$ and $c_t p(Z_t \mid X_t)$ (the density $p(Z_t \mid X_t)$ multiplied by the factor c_t). Following the principle of factored sampling, CONDENSATION generates independent samples from the mixture density $s_t^j \sim \frac{1}{M} \sum_{i=1}^{M} w_{t-1}^i p(X_t \mid X_{t-1} = s_{t-1}^i)$, and assigns weights $p(Z_t \mid X_t = s_t^j)$ (note that c_t is not required in these weights as it is cancelled via normalization). So the method now boils down to the question, *how does one sample from the mixture density*, $\frac{1}{M} \sum_{i=1}^{M} w_{t-1}^i p(X_t \mid X_{t-1} = s_{t-1}^i)$? Since each term $p(X_t \mid X_{t-1} = s_{t-1}^i)$ in this mixture is indeed a density function, sample generation from the mixture may be performed in two steps:

Step 1: Select an index j from the set $\{1, 2, \ldots, M\}$ where the probability of selection for each index is proportional to the corresponding weight in $\{w_{t-1}^1, \ldots, w_{t-1}^M\}$.

Step 2: Generate a sample $s_t^j \sim p(X_t \mid X_{t-1} = s_{t-1}^j)$.

These two steps can be performed M times to generate M independent samples from the mixture density. The CONDENSATION algorithm is illustrated as follows:

$$[\{s_t^i, w_t^i, c_t^i\}_{i=1}^M] = \text{CONDENSATION}[\{s_{t-1}^i, w_{t-1}^i, c_{t-1}^i\}_{i=1}^M]$$

Assign: $c_t^0 \leftarrow 0$

for $i = 1 : M$

      Generate a random number $u \in [0,1]$ uniformly distributed

      Find the smallest j such that $c_{t-1}^j \geq c_{t-1}^M u$

      Generate: $s_t^i \sim p(X_t \mid X_{t-1} = s_{t-1}^j)$

      Assign: $w_t^i \leftarrow p(Z_t \mid X_t = s_t^i)$

      Assign: $c_t^i \leftarrow c_t^{i-1} + w_t^i$

end

The cs in the CONDENSATION algorithm represent the cumulative weights for the samples. The input to the CONDENSATION algorithm is the set $\{(s_{t-1}^1, w_{t-1}^1, c_{t-1}^1), \ldots, (s_{t-1}^M, w_{t-1}^M, c_{t-1}^M)\}$ and the output is $\{(s_t^1, w_t^1, c_t^1), \ldots, (s_t^M, w_t^M, c_t^M)\}$. Since the array of cumulative weights is sorted (in increasing order), the smallest j at the second step inside the loop can be efficiently computed in $O(\log_2 M)$ complexity by a binary search method [19]. Thus the complexity of the CONDENSATION algorithm in terms of sample size M is $O(M \log_2 M)$. We will revisit this algorithm during our case study in Section 4.4.

4.2.2 Auxiliary Particle Filters

Another effective particle filter method is known as the *auxiliary particle filter* (APF) [15,21]. To elucidate the APF we first need to mention the concept of importance sampling [21]. Importance sampling is employed when it is difficult to sample directly from a density $f(x)$. For example the von Mises density is hard to sample directly. In such cases we search for a density function $g(x)$, from which sampling is straightforward. We can then write $f(x)$ as $f(x) = [f(x)/g(x)]g(x)$, where the support of $f(x)$ is a subset of the support of $g(x)$. In other words, $g(x) = 0$ implies $f(x) = 0$. Now we can apply the principle of factored sampling: a set of independent random samples $\{s^1, \ldots, s^M\}$ from the density $g(x)$, along with weights proportional to $\{f(s^1)/g(s^1), \ldots, f(s^M)/g(s^M)\}$, represent the density $f(x)$. The implicit assumption is that one can evaluate both the functions $f(x)$ and $g(x)$ for any x. The density $g(x)$ is referred to as an *importance density*.

To illustrate the APF we rewrite (4.3) as

$$p(X_t \mid Z_{1:t}) \propto \sum_{i=1}^{M} p(X_t, i \mid Z_{1:t}), \tag{4.6}$$

where

$$p(X_t, i \mid Z_{1:t}) = w_{t-1}^i p(Z_t \mid X_t) p(X_t \mid X_{t-1} = s_{t-1}^i). \tag{4.7}$$

The APF principle says that to generate samples from (4.6) one should generate samples of the form $(s, j) \sim p(X_t, i \mid Z_{1:t})$ and then drop the index term j. To generate samples (s, j) from $p(X_t, i \mid Z_{1:t})$, the APF introduces an importance density function as follows:

$$g(X_t, i) \propto w_{t-1}^i p(Z_t \mid X_t = \mu_t^i) p(X_t \mid X_{t-1} = s_{t-1}^i), \tag{4.8}$$

where μ_t^i is the mean (or mode) of the density $p(X_t \mid X_{t-1} = s_{t-1}^i)$, or it may even be a sample from the density $p(X_t \mid X_{t-1} = s_{t-1}^i)$. Note that (4.8) marginalizes as

$$g(i) = \int g(X_t, i) dX_t \propto w_{t-1}^i \int p(Z_t \mid X_t = \mu_t^i) p(X_t \mid X_{t-1} = s_{t-1}^i) dX_t$$
$$= w_{t-1}^i p(Z_t \mid X_t = \mu_t^i), \tag{4.9}$$

so, the conditional density $g(X_t \mid i)$ becomes

$$g(X_t \mid i) = \frac{g(X_t, i)}{g(i)} = \frac{w_{t-1}^i p(Z_t \mid X_t = \mu_t^i) p(X_t \mid X_{t-1} = s_{t-1}^i)}{w_{t-1}^i p(Z_t \mid X_t = \mu_t^i)}$$
$$= p(X_t \mid X_{t-1} = s_{t-1}^i). \tag{4.10}$$

Therefore, the APF generates a sample $(s, j) \sim g(X_t, i)$ by first drawing an index $j \sim g(i)$, where $g(i) \propto w_{t-1}^i p(Z_t \mid X_t = \mu_t^i)$, then drawing a sample $s \sim$

$g(X_t \mid i = j) = p(X_t \mid X_{t-1} = s^j_{t-1})$. Next, following the principle of importance sampling, the weight for the sample (s, j) is set as $w = \frac{p(X_t = s, i = j \mid Z_{1:t})}{g(X_t = s, i = j)} = \frac{p(Z_t \mid X_t = s)}{p(Z_t \mid X_t = \mu^j_t)}$. Finally, the index j is dropped from the generated sample (s, j) and only s is retained. The APF algorithm can be summarized by way of pseudo code:

$$\left[\{s^i_t, w^i_t\}^M_{i=1} \right] = \text{APF} \left[\{s^i_{t-1}, w^i_{t-1}\}^M_{i=1} \right]$$

Assign: $c^0 \leftarrow 0$

for $i = 1 : M$

   Compute: $\mu^i_t \leftarrow \underset{X_t}{\arg \max} \left[p(X_t \mid X_{t-1} = s^i_{t-1}) \right]$

   $c^i \leftarrow c^{i-1} + w^i_{t-1} p(Z_t \mid X_t = \mu^i_t)$

end

for $i = 1 : M$

   Generate a random number $u \in [0, 1]$ distributed uniformly

   Find the smallest j such that $c^j \geq c^M u$

   Draw a sample: $s^i_t \sim p(X_t \mid X_{t-1} = s^j_{t-1})$

   Assign: $w^i_t \leftarrow \frac{p(Z_t \mid X_t = s^i_t)}{p(Z_t \mid X_t = \mu^i_t)}$

end

The input to the APF algorithm is the set of weighted samples $\{(s^1_{t-1}, w^1_{t-1}), \ldots, (s^M_{t-1}, w^M_{t-1})\}$ for the posterior density at the previous time, and the output is the set of weighted samples $\{(s^1_t, w^1_t), \ldots, (s^M_t, w^M_t)\}$ for the current posterior density. In the second step of the first loop of algorithm APF, the $g(i)$ values as in (4.9) are computed and made cumulative. Utilizing the cumulative weights, a sample index j is drawn in the first two steps of the second loop. The third step of the second loop draws a sample from $g(X_t \mid i)$ as in (4.10), and the fourth step of the second loop sets the weight for the sample.

4.3   MULTI-TARGET TRACKING

Biological and medical applications often involve multiple targets. If one were tracking a single cell and had a decent technique in hand, one possible solution for tracking N such cells would be replicating the single-cell technique N times. On the negative side, this replication would lead to an N-fold increase in computational complexity and would not consider cell-to-cell interactions. In terms of tracking lingo, the track from one cell could steal the measurement associated with another cell—or the trackers could be confounded by an apparent merge or split.

To address the issues of tracking multiple targets, we now look at a cornucopia of more sophisticated multi-target tracking (MTT) solutions. MTT is a challenging problem primarily because of the complexity of data association posed by the presence of multiple targets and several measurements. Furthermore, the appearance and disappearance of targets from the field of view beg for a sound technical framework to deal with MTT. By far, the most cultivated framework in MTT is again the sequential Bayesian framework. We have already seen this framework in the Kalman filter and in the particle filter models.

In the next subsections we briefly visit the basic principles of MTT and then describe the Markov chain Monte Carlo (MCMC) method for MTT.

4.3.1   Multiple Hypothesis Tracking

A well-known technical framework for MTT is multiple hypothesis tracking (MHT) [22]. In principle, MHT is not complicated; however, the underlying probabilistic formulation is indeed tedious. Also the computation of MHT is formidable since it increases exponentially with the number of targets. To describe MHT in simple terms, let us define certain necessary terminologies first. A *track* at time t is a target state X_t. In the MTT scenario we have essentially N tracks forming the joint target state $\mathbf{X}_t \equiv \{X_t^1, \ldots, X_t^N\}$. Data association in the MTT context at time t can be deemed a mapping h from the set of detected targets $\{d_1, d_2, \ldots, d_M\}$ to the set of tracks and clutter. In other words,

$$h(d_i) = \begin{cases} j \neq 0 \Rightarrow d_i \text{ is associated with track } j, \\ 0 \Rightarrow d_i \text{ is generated from clutter,} \end{cases} \qquad (4.11)$$

where we assume that a detected target can be associated with at the most one track. We may denote one configuration of the data association hypothesis by a vector $\mathbf{h} \equiv \{h(d_1), \ldots, h(d_M)\}$. MHT computes the posterior distribution of the joint target state $\mathbf{X}_t$ by taking into consideration all possible hypothesis configurations and their associated probabilities:

$$p(\mathbf{X}_t \mid Z_{1:t}) = \sum_{\mathbf{h}} p(\mathbf{h}) p(\mathbf{X}_t \mid \mathbf{h}, Z_{1:t}). \qquad (4.12)$$

MHT computes (4.12) in a sequential fashion. Knowing the previous posterior density of the target tracks, MHT computes the current posterior density from the tracks. The computational complexity is salient from the structure of (4.12), where the summation is taken over all hypothesis configurations. Note that in order to compute (4.12), MHT needs to know the probabilities $p(\mathbf{h})$ for an exponentially growing number of hypothesis configurations. Another subtle point—the detected set of targets $\{d_1, d_2, \ldots, d_M\}$ can actually be composed of targets that belong not only to the current frame t, but also from T future frames $(t + 1), (t + 2), \ldots, (t + T)$. Hence, the computational load of MHT may turn out to be monstrous.

4.3.2 Joint Probabilistic Data Association

Among the approximations of MHT, the joint probabilistic data association (JPDA) method [23] can be a savior. JPDA does not consider future frames of detection (*i.e.*, T is zero in this case). Also, it is possible to prune the total number of viable hypothesis configurations to a significant degree by considering clustering of the tracks based on some features of the target state space (e.g., with respect to some distance measure on the target space). Once such clusters are made, assignment (4.11) will yield fewer configurations, and consequently, the computational complexity of (4.12) is reduced.

Surprisingly, there exist methods unlike MHT and JPDA that avoid explicit techniques for data association. These methods are known as association-free or unified tracking (UT) approaches [22] and are not to be confused with the University of Texas (UT). In MHT, one crucial assumption in the data association hypothesis is that one measurement (*i.e.*, the detected target) can be associated with at most one track. Sometimes one measurement may actually correspond to two or more occluding targets. Consequently, the measurement needs to be associated with two or more tracks. This situation is not permitted under the MHT framework. There are yet other cases where a data association hypothesis is not meaningful. As an example, sometimes a single measurement is taken for all tracks. In such a case data association is not meaningful, and consequently, the MHT approach is not suitable for these applications. The UT method comes to the rescue in such cases. (Aside: MHT is a special case of the unified tracking approach [22].)

The state estimation recursions for the UT approach in the multi-target scenario are similar to the ones in the single target scenario

$$p(\mathbf{X}_t \mid Z_{1:t}) = \frac{p(Z_t \mid \mathbf{X}_t)p(\mathbf{X}_t \mid Z_{1:t-1})}{\int p(Z_t \mid \mathbf{X}_t)p(\mathbf{X}_t \mid Z_{1:t-1})d\mathbf{X}_t}, \qquad (4.13)$$

where $p(\mathbf{X}_t \mid Z_{1:t-1})$ is a joint prior density for all the tracks and is defined as follows:

$$p(\mathbf{X}_t \mid Z_{1:t-1}) = \int p(\mathbf{X}_t \mid \mathbf{X}_{t-1})p(\mathbf{X}_{t-1} \mid Z_{1:t-1})d\mathbf{X}_{t-1}. \qquad (4.14)$$

Compare Eqs. (3.5) and (3.6) with Eqs. (4.13) and (4.14). Similar to the single target case, we have the following representation for the joint posterior density for all N targets:

$$p(\mathbf{X}_t \mid Z_{1:t}) \propto p(Z_t \mid \mathbf{X}_t) \sum_{i=1}^{M} w_{t-1}^i p(\mathbf{X}_t \mid \mathbf{X}_{t-1} = \mathbf{s}_{t-1}^i), \qquad (4.15)$$

where $\mathbf{s}_{t-1}^i \equiv \{s_{t-1}^{j,i}\}_{j=1}^{j=N}$ is a sample from the joint posterior density $p(\mathbf{X}_{t-1} \mid Z_{1:t-1})$. Again the task is to generate samples $\{\mathbf{s}_t^i\}_{i=1}^{M} \equiv \{s_t^{j,i}\}_{j=1,i=1}^{j=N,i=M}$ for the N tracks to approximate the current joint posterior $p(\mathbf{X}_t \mid Z_{1:t})$. We accomplish this sampling by way of a Markov chain Monte Carlo (MCMC) method [24].

4.3.3   Markov Chain Monte Carlo Methods

A few specialized terms from the Markov world are prerequisite for the illustration of the MCMC algorithm. Let us designate the *state* of a system at *pseudo*-time i by S^i. The state S^i is a random variable, and let us assume for the sake of simplicity that it takes values from a finite set. We will denote an instance (value) of the random variable S^i by s^i. The instance s^i is a member of the finite set of states. In the MTT scenario, the state essentially represents the positions and/or velocities of targets in an image sequence.

The system is called *first order* Markov when the probability of the system being at the current state is conditionally independent given the immediate previous state of the system—the current state depends only on the previous state. For example, if the system encompasses the states $s^0, s^1, s^2, \ldots, s^i$, at pseudo-times 0, 1, 2, ..., i, respectively, then $p(S^{i+1} \mid S^i = s^i, \ldots, S^0 = s^0) = p(S^{i+1} \mid S^i = s^i)$. For such a system, the states $s^0, s^1, s^2, \ldots, s^i$ are said to form a Markov chain. Markov chain systems can be described by (a) the probabilities $p(S^0)$ for all possible initial states and (b) the state transition probabilities for all possible pairs of states, as the probability of the state at $(i + 1)$th pseudo-time instant is given by

$$p(S^{i+1}) = \sum_{s^i} p(S^i)p(S^{i+1} \mid S^i). \tag{4.16}$$

In general $p(S^{i+1}) \neq p(S^i)$ for any pseudo-time i; however, when this equality holds for all states, the probability $p(\cdot)$ is deemed the stationary distribution $\pi(\cdot)$ of the Markov chain.

4.3.4   Metropolis–Hastings Sampling

A widely applied MCMC sampling algorithm in the biomedical image analysis community is Metropolis–Hastings (MH) sampling. The MH method sequentially generates states $s^0, s^1, \ldots, s^i$, which is a realization of a Markov chain. After simulating the Markov chain for a sufficient time, the stationary distribution is reached for the Markov chain. Let us assume that the MH algorithm has already generated a Markov chain $s^1, s^2, \ldots, s^i$. To generate the $(i + 1)$th member in the chain, the algorithm first generates a sample $s \sim q$, where $q(S^{i+1}; S^i)$ is known as the proposal density that generates a sample s for the $(i + 1)$th state S^{i+1}, while the chain is currently at state $S^i = s^i$. Then the MH algorithm accepts or rejects the sample s according to the following threshold:

$$r = \frac{\pi(s)q(s^i; s)}{\pi(s^i)q(s; s^i)}. \tag{4.17}$$

A uniformly distributed random number $u \in [0,1]$ is drawn. If $r \geq u$ then the sample s is accepted (*i.e.*, s^{i+1} is set as s). If $r < u$ the sample s is rejected, and the last state sample is repeated (*i.e.*, s^{i+1} is set as s^i). The choice of the proposal distribution q is problem-dependent, and a proper choice for q yields fast convergence to stationary distribution. The samples $s^1, s^2, \ldots, s^i, \ldots$ are correlated and not independent; however, they serve to estimate the expected value of a function $h(X)$ of the random variable X as follows [24]:

$$\int h(X)\pi(X)dX \approx \frac{1}{N} \sum_{i=1}^{N} h(s^i). \tag{4.18}$$

The accuracy of the estimate (4.18) increases as N increases. In biomedical target tracking we are often interested in the expected value of the states. So once a chain of samples is generated via the MH algorithm, we can find the expected value of the state by simply substituting s^i for $h(s^i)$ in (4.18).

To turn our attention back to the MTT problem, let us assume that at the $(t - 1)$th frame, we have generated M samples $\{\mathbf{s}_{t-1}^i\}_{i=1}^{M} \equiv \{s_{t-1}^{j,i}\}_{j=1,i=1}^{j=N,i=M}$ which form a Markov chain. By the law of large numbers for Markov chain (4.15) can be represented as

$$p(\mathbf{X}_t \mid Z_{1:t}) = c_t p(Z_t \mid \mathbf{X}_t) \sum_{i=1}^{M} p(\mathbf{X}_t \mid \mathbf{X}_{t-1} = \mathbf{s}_{t-1}^i). \tag{4.19}$$

Once again c_t is a normalization factor never required in the MH algorithm. To make the recursion in (4.19) come alive, the MCMC algorithm now generates samples $\{\mathbf{s}_t^i\}_{i=1}^M \equiv \{s_t^{j,i}\}_{j=1,i=1}^{j=N,i=M}$ to simulate another Markov chain, the stationary distribution, which is characterized by the current posterior density $p(\mathbf{X}_t \mid Z_{1:t})$. Note that each state $\mathbf{s}^i$ has N components, because we are dealing with N targets.

When the states of a Markov chain are multi-component, MH provides a straightforward way to generate sample vectors $s^i \equiv \{\mathbf{s}^{j,i}\}_{j=1}^N$: the ratio (4.17) can be computed for a single component $s^{j,i}$ as opposed to the entire sample vector $\mathbf{s}^i$. The component can be rejected/accepted in the same way as described before. This particular feature of the MH algorithm is helpful for the MTT problem. The MH algorithm for MTT must sample from the mixture density (4.19). MH first randomly selects the kth component of the mixture density (4.19). The MH approach then generates a (single component) sample state for the jth target X_t^j using the following proposal density:

$$q(X_t^j) = p(X_t^j \mid \{X_t^l : l \neq j\}, \mathbf{X}_{t-1} = \mathbf{s}_{t-1}^k). \tag{4.20}$$

Note that the motion model $p(\mathbf{X}_t \mid \mathbf{X}_{t-1})$ factors as

$$p(\mathbf{X}_t \mid \mathbf{X}_{t-1} = \mathbf{s}_{t-1}^k) = p(X_t^j \mid \{X_t^l : l \neq j\},$$
$$\mathbf{X}_{t-1} = \mathbf{s}_{t-1}^k)p(\{X_t^l : l \neq j\}, \mathbf{X}_{t-1} = \mathbf{s}_{t-1}^k) \text{ for any } j. \tag{4.21}$$

Thus the MH ratio (4.17) for the jth component takes the following form:

$$r = \frac{p(Z_t \mid X^j = s^{\text{new}}, \{X_t^l : l \neq j\})p(X^j = s^{\text{new}}, \{X_t^l : l \neq j\} \mid \mathbf{X}_{t-1} = \mathbf{s}_{t-1}^k)q(X_t^j = s)}{p(Z_t \mid X^j = s, \{X_t^l : l \neq j\})p(X^j = s, \{X_t^l : l \neq j\} \mid \mathbf{X}_{t-1} = \mathbf{s}_{t-1}^k)q(X_t^j = s^{\text{new}})}$$
$$= \frac{p(Z_t \mid X^j = s^{\text{new}}, \{X_t^l : l \neq j\})}{p(Z_t \mid X^j = s, \{X_t^l : l \neq j\})}, \tag{4.22}$$

where s^{new} is a sample drawn from the proposal distribution (4.20) and s is the current jth target state. The second equality in (4.22) is obtained by utilizing

(4.20) and (4.21). Having mentioned the prerequisites about MCMC and Markov chains, we now outline a generic MCMC algorithm for the MTT problem:

$$\{s_t^{i,j}\}_{j=1,i=1}^{j=N,i=M} = \text{MTT\_MCMC}\left[\{S_{t-1}^{i,j}\}_{j=1,i=1}^{j=N,i=M}\right]$$

Choose initial samples $\{s_t^{j,0}\}_{j=1}^{j=N}$

for $i = 1 : M$

    Choose an index $k \in \{1, \ldots, M\}$, distributed uniformly

    for $j = 1 : N$

        Draw: $s \sim p\left(X_t^j \mid X_t^1 = s_t^{1,i}, \ldots, X_t^{j-1} = s_t^{j-1,i}, X_t^{j+1} = s_t^{j+1,i-1}, \ldots, X_t^N = s_t^{N,i-1}, X_{t-1}^1 = s_{t-1}^{1,k}, \cdots, X_{t-1}^N = s_{t-1}^{N,k}\right)$

        Compute:

$$r = \frac{p\left(Z_t \mid X_t^1 = s_t^{1,i}, \ldots, X_t^{j-1} = s_t^{j-1,i}, X_t^j = s, X_t^{j+1} = s_t^{j+1,i-1}, \ldots, X_t^N = s_t^{N,i-1}\right)}{p\left(Z_t \mid X_t^1 = s_t^{1,i}, \ldots, X_t^{j-1} = s_t^{j-1,i}, X_t^j = s_t^{j,i-1}, X_t^{j+1} = s_t^{j+1,i-1}, \ldots, X_t^N = s_t^{N,i-1}\right)}$$

        Choose: $u \in [0,1]$, uniformly distributed

        if $r \geq u$

            Assign: $s_t^{j,i} \leftarrow s$

        else

            Assign: $s_t^{j,i} \leftarrow s_t^{j,i-1}$

        end

    end

end

Note that the proposal density for a single target state is (4.20) in the algorithm MTT_MCMC. Equation (4.20) is essentially the conditional motion model. Alternatively, if we assume that motion of each target is conditionally independent on the previous target state, then the proposal (4.20) takes the following form:

$$q(X_t^j) = p(X_t^j \mid X_{t-1}^j = s_{t-1}^{j,k}). \tag{4.23}$$

The algorithm MTT_MCMC utilizes the ratio r from (4.22) to accept/reject a proposal state sampled from (4.20).

4.3.5 Auction Algorithm

In biomedical tracking applications, correspondence resolution can be more troublesome than object detection. An example is *in vitro* video microscopy with congested populations of cells, microbubbles (small bubbles used as contrast agents and for drug delivery), and/or other components. Given a set of detected cells within a video frame and another set of detected cells from a subsequent frame, we want to resolve the correspondence or equivalently establish a mapping from one set to the other. The aim of this section is to examine a clever solution to the correspondence problem, called the *auction algorithm* [25].

Let us first consider the one-to-one assignment problem. Here, there are n *people* and n *objects*. We want to match people and objects on a one-to-one basis. Let a_{ij} denote the benefit of matching the ith person with the jth object. Following [25], the problem is defined as one of maximizing *benefit*. We want to find a set of person–object pairs $(1, j(1)), (2, j(2)), \ldots, (n, j(n))$ such that the distinct objects $j(1), j(2), \ldots, j(n)$ maximize the total benefit $\sum_{i=1}^{n} a_{ij(i)}$.

How can one solve this benefit-maximization problem? If you have seen the movie *A Beautiful Mind*, you might be able to guess ... According to [25], it turns out that if one solves an equilibrium problem from economics, the assignment problem is also solved. *So what is this equilibrium problem?* Referring back to the people and the objects, let us assume that each person is acting in his/her best interest and is contending for an object. However each object j has a price tag p_j. Thus the net value of object j for person i is $(a_{ij} - p_j)$. Each person i wants an object j_i, such that the net value of the object is maximized, *i.e.*,

$$a_{ij(i)} - p_{j_i} = \max_{j=1,\ldots,n} \{a_{ij} - p_j\}. \tag{4.24}$$

We say that the one-to-one assignment and object prices are at equilibrium when condition (4.24) holds for all persons $i = 1, \ldots, n$. As already stated, if (4.24) holds for all of the people, then we have solved the assignment problem.

An auction process is a natural way to achieve this equilibrium [25]. The auction starts with any set of prices and any one-to-one assignment between people and objects. The process terminates when equilibrium is reached; otherwise it iterates as follows:

Step 1: Find a person i for whom condition (4.24) does not hold.

Step 2: Compute $j(i) = \arg\max_{j=1,\ldots,n}\{a_{ij} - p_j\}$.

Step 3: Find the person k who is assigned object $j(i)$; then let person k and person i exchange their objects.

Step 4: Compute $v_i = \max_{j=1,\ldots,n}\{a_{ij} - p_j\}$.

Step 5: Compute $w_i = \max_{j\in\{1,\ldots,n\}\setminus\{j(i)\}}\{a_{ij} - p_j\}$.

Step 6: Set $p_{j(i)} \leftarrow p_{j(i)} + v_i - w_i$.

Step 7: Check if condition (4.24) holds for all persons. If yes, terminate the process; else, repeat Steps 1 through 7.

Note that v_i is the best object value, and w_i is the second best object value for person i. The bidder i (*i.e.*, person i) raises the price of his/her object of interest by the amount $(v_i - w_i)$, so that other bidders become less interested in this particular object. *Is this auction algorithm guaranteed to reach equilibrium?* Unfortunately the answer is *no* [25]. The algorithm may become stuck in an endless loop ... Note that $v_i = w_i$, when more than one object offers maximum value for person i. In such a case, the price p_{j_i} in Step 7 does not increase. Now, if a situation arises where several persons contend for a set of equally desirable objects, then no one will increase the price of any object, and the algorithm falls in an endless loop. One way to avoid falling into this endless loop is to increase the price of an object at least by a small increment in each bid. This strategy is actually adopted from the real auction process [25]. We define this real auction algorithm next.

Instead of condition (4.24) let us consider the following:

$$a_{ij(i)} - p_{j(i)} \geq \max_{j=1,\ldots,n}\{a_{ij} - p_j\} - \varepsilon, \qquad (4.25)$$

where ε is a small positive number. When condition (4.25) holds for each person i, the process is said to be "almost at equilibrium." Now, as with a real auction process, the algorithm raises the price of an object by at least ε. So Step 6 in the algorithm is modified as "Set $p_{j(i)} \leftarrow p_{j(i)} + v_i - w_i + \varepsilon$." With this modification the algorithm is guaranteed to terminate. Moreover, upon termination it can be shown that the algorithm solves the assignment problem (or maximizes the total benefit $\sum_{i=1}^{n} a_{ij_i}$) if

(a) $\varepsilon < 1/n$, and

(b) a_{ij} are all integers.

If a_{ij} are rational numbers, then we can suitably scale up a_{ij} to integers by multiplying them by a suitable number; the auction algorithm will work just as well.

For the correspondence resolution problem in the case of MTT, this algorithm must be modified, because the number of detected targets (such as cells) on two consecutive video frames differs. The good news is that with only a suitable initial condition in the algorithm, this asymmetric assignment problem can be solved. Without loss of generality, let us assume that the number of detected objects in the previous frame is greater than that in the current frame. Further we require that each detected object on the current frame should be assigned an object in the previous frame. In this case, if the algorithm starts with initial price of zero for all the objects, then the auction algorithm outlined applies without any further modification. In tracking, multiple assignments may be required instead of the one-to-one assignment. For example, often in biomedical applications, two or more objects merge into a single object or a single object splits into two or more objects. Suitable modifications of the basic auction algorithm exist in such cases as well.

4.4   CASE STUDIES

We provide two case studies in this section in order to demonstrate the application of the CONDENSATION and MCMC sampling methods:

1. *Single target tracking using the CONDENSATION algorithm for non-Gaussian, nonlinear models.* The application is tracking cells, specifically rolling leukocytes observed within intravital video microscopy.

2. *Multiple target tracking by way of Markov chain Monte Carlo sampling (MTT_MCMC)* to track multiple cells from an *in vitro* microscopy assay.

The purpose of the case studies is to illustrate how the probabilistic theory of tracking is applied in practice to problems that we frequently encounter in biomedical applications.

4.4.1   Leukocyte Tracking with CONDENSATION

4.4.1.1 Polar Model

Let us now consider a non-Gaussian motion model and a nonlinear measurement model in tracking rolling leukocytes from *in vivo* microscopy (for a detailed description of the application, see Chapter 2, Section 2.4). We convert the Cartesian coordinates (x_t, y_t) for the cell location into polar coordinates (r_t, τ_t) and express the motion model in the polar coordinates as follows:

$$p(r_t, \tau_t \mid r_{t-1}, \tau_{t-1}) = \frac{1}{R} \frac{\exp\left[\sigma_v \cos(\tau_t - \tau_{t-1})\right]}{2\pi I_0(\sigma)}, \tag{4.26}$$

where $r_t \in [r_{t-1}, r_{t-1} + R]$, $\tau_t \in [0, 2\pi]$, and $I_0(\cdot)$ is a modified Bessel function of the first kind and is given as

$$I_0(\sigma) = \sum_{k=0}^{\infty} \frac{(\sigma^2/4)^2}{(k!)^2}. \tag{4.27}$$

The polar motion model (4.26) has a simple interpretation in which the speed of the cell is uniformly distributed and the direction of movement is distributed as a von Mises distribution with parameter σ [26].

4.4.1.2 Measurement Model: Gradient Inverse Coefficient of Variation

To define the measurement probability model $p(Z_t \mid X_t)$, let us first define a measurement statistic—the gradient inverse coefficient of variation (GICOV). The GICOV is the ratio of the mean and the standard deviation of the image intensity directional derivatives, computed over the length of a predefined curve. Since the

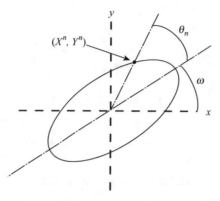

FIGURE 4.1: An oriented ellipse and a point on the ellipse. The center of the ellipse is the origin.

leukocytes appear approximately elliptical in the *in vivo* imagery, we restrict these curves to be ellipses (see Fig. 4.1). An ellipse can be expressed by five parameters: semi-major axis length (a), semi-minor axis length (b), orientation (ω), and center coordinates (x, y). Furthermore, to compute the GICOV statistic, we need N discrete points on the ellipse. We discretize the ellipse using N ellipse points in discrete steps at angles $\theta_n = (2\pi n/N)$, $n = 0, \ldots, N-1$ with respect to the center of the ellipse. The nth point on the ellipse with five parameter values (a, b, ω, x, y) has coordinates (see Fig. 4.1)

$$(X^n, Y^n) \equiv [x + a\cos(\theta_n)\cos(\omega) - b\sin(\theta_n)\sin(\omega), \, y + a\cos(\theta_n)\sin(\omega)$$
$$+ b\sin(\theta_n)\cos(\omega)].$$

To be technically precise, the GICOV is the ratio of the sample mean over the normalized sample standard deviation $\frac{m(a,b,\omega,x,y)}{S(a,b,\omega,x,y)/\sqrt{N}}$, where the sample mean $m(\cdot)$ and the sample standard deviation $S(\cdot)$ are defined respectively as

$$m(a, b, \omega, x, y) = \frac{1}{N}\sum_{n=1}^{N}\left(I\left(X^n + \frac{dX^n}{2}, Y^n + \frac{dY^n}{2}\right)\right.$$
$$\left. - I\left(X^n - \frac{dX^n}{2}, Y^n - \frac{dY^n}{2}\right)\right), \tag{4.28}$$

and via

$$S^2(a, b, \omega, x, y) = \frac{1}{N-1} \sum_{n=1}^{N} \left(I\left(X^n + \frac{dX^n}{2}, Y^n + \frac{dY^n}{2} \right) \right.$$

$$\left. - I\left(X^n - \frac{dX^n}{2}, Y^n - \frac{dY^n}{2} \right) \right)^2$$

$$- \frac{N}{N-1} m^2(a, b, \omega, x, y), \tag{4.29}$$

where I is the associated matrix of image intensities, and (dX^n, dY^n) denotes the unit outward normal at the nth ellipse point and can be explicitly defined as

$$(dX^n, dY^n) \equiv \left[\frac{b \cos(\theta_n) \cos(\omega) - a \sin(\theta_n) \sin(\omega)}{\sqrt{a^2 \sin^2(\theta_n) + b^2 \cos^2(\theta_n)}}, \right.$$

$$\left. \times \frac{b \cos(\theta_n) \sin(\omega) + a \sin(\theta_n) \cos(\omega)}{\sqrt{a^2 \sin^2(\theta_n) + b^2 \cos^2(\theta_n)}} \right]. \tag{4.30}$$

Figure 4.2 illustrates the positioning of $(X^n + (dX^n/2), Y^n + (dY^n/2))$ and $(X^n + (dX^n/2), Y^n + (dY^n/2))$.

Let us assume that a constant-intensity image is corrupted with zero mean Gaussian noise of variance σ^2, and that the leukocytes are objects with step edge of intensity height A over this constant intensity image. If we further assume that an ellipse delineates a cell, then it can be shown following the analysis in [27] that the distribution of the associated GICOV statistic $\frac{m(a,b,\omega,x,y)}{S(a,b,\omega,x,y)/\sqrt{N}}$ is a noncentral

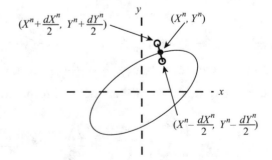

FIGURE 4.2: The points (X^n, Y^n), $(X^n + (dX^n/2), Y^n + (dY^n/2))$, and $(X^n - (dX^n/2), Y^n - (dY^n/2))$ on an ellipse.

Student's t distribution with a noncentrality parameter $\sqrt{N/2}(A/\sigma)$ and $N-1$ degrees of freedom. So the question of how to compute the five ellipse parameter values (a, b, ω, x, and y) now arises. We hypothesize that if an ellipse center is the cell center, then taking the maximum of $\frac{m(a,b,\omega,x,y)}{S(a,b,\omega,x,y)/\sqrt{N}}$ over a range of the three parameter values (a, b, and ω), we can compute the maximum GICOV value that corresponds to an ellipse delineating a cell. Therefore the maximum GICOV value, given by

$$g(x, y) = \max_{a,b,\theta} \left(\frac{m(a, b, \omega, x, y)}{S(a, b, \omega, x, y)/\sqrt{N}} \right), \tag{4.31}$$

is noncentral Student's t-distributed, with noncentrality parameter $\sqrt{N/2}(A/\sigma)$, and $N-1$ degrees of freedom. For a faster computation, the noncentral Student's t density can be approximated by a Gaussian distribution. In this case, the statistic

$$\gamma(x, y) = \frac{g(x, y)\left[1 - \dfrac{1}{4(N-1)}\right] - \sqrt{\dfrac{N}{2}}\dfrac{A}{\sigma}}{\left[1 + \dfrac{g^2(x, y)}{2(N-1)}\right]^{1/2}}, \tag{4.32}$$

is Gaussian distributed, with a mean of zero and variance of one [28].

We are now finally ready to spell out the measurement model for this cell tracking problem:

$$p(Z_t \mid x_t, y_t) \propto \exp\left[-\frac{\gamma^2(x_t, y_t)}{2}\right]. \tag{4.33}$$

Note that in order to utilize the measurement model (4.23), we need to estimate the noise standard deviation σ, and the leukocyte step edge height A from a set of training images. The estimate for σ can be computed from an arbitrarily selected homogenous region in the image. To estimate A, we manually detect a few leukocytes on the first video frame and compute the mean step edge strength.

Notice that in order to implement sequential Bayesian tracking, we need three probability densities, $viz.$, $p(X_0)$, $p(X_t \mid X_{t-1})$, and $p(Z_t \mid X_t)$. We have provided solutions for $p(X_t \mid X_{t-1})$ and $p(Z_t \mid X_t)$ using (4.26) and (4.33), respectively. The initial state density $p(X_0)$ can be designed as a Kroneker delta function:

$$p(x_0, y_0) \equiv \begin{cases} 1, & \text{if } x_0 = x_u \text{ and } y_0 = y_u, \\ 0, & \text{otherwise.} \end{cases} \tag{4.34}$$

where the (x_u, y_u) is the manually determined leukocyte center on the first video frame.

The leukocyte tracking algorithm via CONDENSATION is provided in the algorithmic form and labeled as SINGLE_LEUKO_TRACK. The algorithm shown is for the (t)th image frame. To initialize the algorithm, we set $t \leftarrow 1$ and $(x_0^i, y_0^i) \leftarrow (x_u, y_u)$, $w_0^i \leftarrow 1/M$, and $c_0^i \leftarrow i/M$, $for\ i = 1, \ldots, M$. Note that to begin tracking, the user chooses a leukocyte center (x_u, y_u) on the initial (0th) frame. Automated tracking through the algorithm SINGLE_LEUKO_TRACK begins on frame 1. The algorithm outputs the estimated leukocyte center $(\bar{x}_t, \bar{y}_t)$ for every frame t.

There are a few parameters that we need to set beforehand for this algorithm. We enumerate them here: (1) the speed range R as in (4.26), (2) the von Mises parameter σ_v as in (4.26), (3) the range of values for the ellipse parameters ($a, b,$ and ω), (4) the number of discrete steps within the range of values for the parameters ($a, b,$ and ω), (5) the number of ellipse sample points N, and (6) the number of samples M in the tracking algorithm. Some of these parameters can be chosen by way of domain knowledge. For example, since the approximate size of a rolling leukocyte is known, we can accordingly set a range of semi-major and semi-minor axes a and b for the ellipses. The range for ellipse orientation ω can be chosen from extensive observations and knowledge of the vascular flow. The number of discrete steps to be considered within these ranges can also be chosen on the basis of the performance of cell detection on a set of training images. The speed range R may also come from the domain knowledge about the maximum speed of rolling leukocytes. Choosing the von Mises parameter is difficult—we do it here on the basis of *ad hoc* visual perception of the tracking performance.

The algorithm is described in pseudo code as

$$\left[(\bar{x}_t, \bar{y}_t), \{(x_t^i, y_t^i), w_t^i, c_t^i\}_{i=1}^M\right]$$
$$= \text{SINGLE\_LEUKO\_TRACK}\left[\{(x_{t-1}^i, y_{t-1}^i), w_{t-1}^i, c_{t-1}^i\}_{i=1}^M\right]$$

Assign: $c_t^0 \leftarrow 0$

for $i = 1 : M$

 Generate a random number $u \in [0,1]$ uniformly distributed

 Find the smallest j such that $c_{t-1}^j \geq c_{t-1}^M u$

 Convert Cartesian (x_{t-1}^j, y_{t-1}^j) to polar $(r_{t-1}^j, \tau_{t-1}^j)$

 Generate: $(r_t^i, \tau_t^i) \sim p(r_t, \tau_t \mid r_{t-1}^j, \tau_{t-1}^j)$ /* equation (4.26) */

 Convert polar (r_t^i, τ_t^i) to Cartesian (x_t^i, y_t^i)

 Compute maximum GICOV: $g(x_t^i, y_t^i)$ /* equation (4.31) */

 Compute: $\gamma(x_t^i, y_t^i)$ /* equation (4.32) */

 Assign: $w_t^i \leftarrow p(Z_t \mid x_t^i, y_t^i)$ /* equation (4.33) */

 Assign: $c_t^i \leftarrow c_t^{i-1} + w_t^i$

end

 Estimate Cell Center: $\bar{x}_t \leftarrow \dfrac{1}{c_t^M} \sum_{i=1}^M x_t^i w_t^i, \ \bar{y}_t \leftarrow \dfrac{1}{c_t^M} \sum_{i=1}^M y_t^i w_t^i.$

Figure 4.3 illustrates several frames of tracking by the aforementioned method. Although the frames show multiple leukocytes being tracked, each cell is tracked individually.

4.4.2 Multiple Cell Tracking with MCMC

In this section we show how the MTT_MCMC algorithm for tracking multiple targets can be applied to track cells from *in vitro* video microscopy. In order to perform an expedited computation, we first detect cell centers within each video frame. For detection of the cells used in this example (rolling leukocytes), we employ morphological operations followed by gray level thresholding. Next, for each target j, once again we express the motion model in polar coordinates:

$$p(r_t, \tau_t \mid r_{t-1}, \tau_{t-1}) \propto \exp\left[-\frac{(r_t - r_{t-1})^2}{2\sigma_r}\right] \exp\left[-\frac{(\tau_t - \tau_{t-1})^2}{2\sigma_\tau}\right], \qquad (4.35)$$

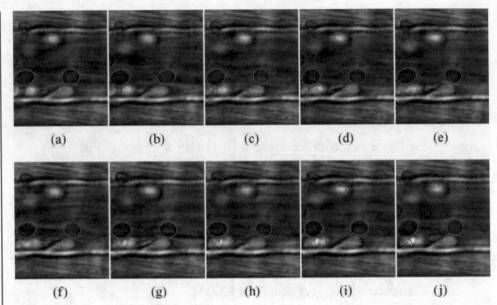

FIGURE 4.3: Tracking leukocytes with the particle filter. The center of the ellipse is computed via CONDENSATION. Each ellipse shown here corresponds to the best GICOV score yielded by an ellipse from a set of ellipses within a small database of ellipses with different orientations and radii. Ten consecutive frames of the sequence are shown here.

where σ_r and σ_τ are standard deviations. For adherent cells, the model of (4.35) is appropriate. A more general motion model will actually allow interactions among targets.

The algorithm MULTI_LEUKO_TRACK chooses a sample $(x_{t-1}^{j,k}, y_{t-1}^{j,k})$ for target j from frame $t-1$ and converts this to polar coordinates $(r_{t-1}^{j,k}, \tau_{t-1}^{j,k})$. Then based on this sample, a set G_j for target j is computed. The set G_j is basically the set of detected targets that fall within a circular window centered at $(x_{t-1}^{j,k}, y_{t-1}^{j,k})$. The radius of the window is based on the maximum movement of each leukocyte in the *in vitro* assay. The next task in the algorithm is to draw samples from the proposal density. We design the proposal probability mass function in terms of the motion model (4.35) and the set G_j. One way to construct this probability mass function is to use the weighted set of coordinates: $\{(r^l, \tau^l), p(r^l, \tau^l \mid r_{t-1}^{j,k}, \tau_{t-1}^{j,k})\}$ for all the elements l in the set G_j. We can then draw a sample from this weighted set, treating the weights $p(r^l, \tau^l \mid r_{t-1}^{j,k}, \tau_{t-1}^{j,k})$ as proportional to the probability of a sample (r^l, τ^l).

Once a sample is chosen, the next task is to compute the Metropolis–Hastings ratio, and either accept or reject the sample as described earlier in the general MTT_MCMC algorithm. We now need the measurement probability model. We will consider a joint measurement probability for the targets:

$$p\left(Z_t \mid \{(x_t^j, y_t^j)\}_{j=1}^M\right) \propto \prod_{x,y \in T_{in}} \exp\left\{-[I_t(x, y) - \mu_{in}]^2 / 2\sigma_{in}^2\right\}$$

$$\times \prod_{x,y \in T_{out}} \exp\left\{-[I_t(x, y) - \mu_{out}]^2 / 2\sigma_{out}^2\right\}, \quad (4.36)$$

where T_{in} is the set of pixels belonging to the region covered by all the targets $\{(x_t^j, y_t^j)\}_{j=1}^N$, and T_{out} is the set of pixels belonging to the outside of all the targets. Thus, T_{in} and T_{out} are disjoint sets and their union yields the entire observation area of the rectangular image domain. Parameters μ_{in} and σ_{in} are the mean intensity and the standard deviation, respectively, inside the leukocytes and similarly μ_{out} and σ_{out} are those for the background of the image. Considering the fact that T_{in} and T_{out} are disjoint sets and that their union is the entire image domain (which is fixed for every frame t), the observation/measurement density (4.36) can also be written as

$$p(Z_t \mid \{(x_t^j, y_t^j)\}_{j=1}^N) \propto \prod_{x,y \in T_{in}} \frac{\exp\left\{-[I(x, y) - \mu_{in}]^2 / 2\sigma_{in}^2\right\}}{\exp\left\{-[I(x, y) - \mu_{out}]^2 / 2\sigma_{out}^2\right\}}. \quad (4.37)$$

So, the Metropolis–Hastings ratio can be written as

$$R = \frac{\displaystyle\prod_{x,y \in T'_{in}} \frac{\exp\left\{-[I(x, y) - \mu_{in}]^2 / 2\sigma_{in}^2\right\}}{\exp\left\{-[I(x, y) - \mu_{out}]^2 / 2\sigma_{out}^2\right\}}}{\displaystyle\prod_{x,y \in T_{in}} \frac{\exp\left\{-[I(x, y) - \mu_{in}]^2 / 2\sigma_{in}^2\right\}}{\exp\left\{-[I(x, y) - \mu_{out}]^2 / 2\sigma_{out}^2\right\}}}, \quad (4.38)$$

where T_{in} denotes the region covered by samples from the previous step, and T'_{in} denotes the region covered by the current samples. Equation (4.38) can be further simplified into

$$R = \frac{\displaystyle\prod_{x,y \in T'_{\text{in}} \backslash (T'_{\text{in}} \cap T_{\text{in}})} \frac{\exp\left\{-[I(x,y) - \mu_{\text{in}}]^2/2\sigma_{\text{in}}^2\right\}}{\exp\left\{-[I(x,y) - \mu_{\text{out}}]^2/2\sigma_{\text{out}}^2\right\}}}{\displaystyle\prod_{x,y \in T_{\text{in}} \backslash (T'_{\text{in}} \cap T_{\text{in}})} \frac{\exp\left\{-[[I(x,y) - \mu_{\text{in}}]]^2/2\sigma_{\text{in}}^2\right\}}{\exp\left\{-[I(x,y) - \mu_{\text{out}}]^2/2\sigma_{\text{out}}^2\right\}}}, \tag{4.39}$$

since the contributions due to the set $T'_{\text{in}} \cap T_{\text{in}}$ cancel within the numerator and the denominator. Note that the only change in T'_{in} from T_{in} will be observed due to a single target; so the cardinality of the set $T'_{\text{in}} \cap T_{in}$ is much larger compared to the that of the two sets $T'_{\text{in}} \backslash (T'_{\text{in}} \cap T_{\text{in}})$ and $T_{\text{in}} \backslash (T'_{\text{in}} \cap T_{\text{in}})$. Thus it is computationally less expensive to compute (4.39), when compared to computing (4.38) in MULTI_LEUKO_TRACK. Also, note that while choosing index k we restricted the set to $\{\lceil 0.2M \rceil, \lceil 0.2M \rceil + 1, \ldots, M\}$, rather than choosing it from the set 1 through M. In fact, we are discarding first 20% of the previous samples. The reason for this is that the MCMC chain typically needs a bit of a "burn in" period in the MCMC algorithm; usually the first 10–20% samples are highly correlated and discarded before accepting samples for computation. For the same reason, we also discard the first 20% samples while estimating cell centers via the mode operation.

In pseudocode, this algorithm is outlined as

$$\left\{(\bar{x}_t^j, \bar{y}_t^j), \{(x_t^{i,j}, y_t^{i,j})\}_{i=1}^M\right\}_{j=1}^N$$
$$= \text{MULTI\_LEUKO\_TRACK}\left[\{\{(x_t^{i,j}, y_t^{i,j})\}_{i=1}^M\}_{j=1}^N\right]$$

Choose initial samples $\left\{(x_t^{j,0}, y_t^{j,0})\right\}_{j=1}^{j=N}$

for $i = 1 : M$

    Choose an index $k \in \{\lceil 0.2M \rceil, \lceil 0.2M \rceil + 1, \ldots, M\}$, distributed uniformly

    Convert Cartesian $(x_{t-1}^{j,k}, y_{t-1}^{j,k})$ to polar $(r_{t-1}^{j,k}, \tau_{t-1}^{j,k})$

    for $j = 1 : N$

        Determine the gate G_j

        Draw: $(r, \tau) \sim \{(r^l, \tau^l), p(r^l, \tau^l \mid r_{t-1}^{j,k}, \tau_{t-1}^{j,k})\}_{l \in G_j}$

        Convert polar (r, τ) to Cartesian (x, y)

        Compute Metropolis–Hastings ratio R in (4.39)

Choose: $u \in [0, 1]$, uniformly distributed

if $R \geq u$

    Assign: $(x_t^{j,i}, y_t^{j,i}) \leftarrow (x, y)$

else

    Assign: $(x_t^{j,i}, y_t^{j,i}) \leftarrow (x_t^{j,i-1}, y_t^{j,i-1})$

end

  end

end

for $j = 1 : N$

    Compute mode: $(\bar{x}_t^j, \bar{y}_t^j) \leftarrow \text{mode}\left[\left\{(x_t^{i,j}, y_t^{i,j})\right\}_{i=\lceil 0.2M \rceil}^{M}\right]$ for $j = 1 \ldots N$

end

Figure 4.4 illustrates two consecutive frames of tracking cells observed via video microscopy.

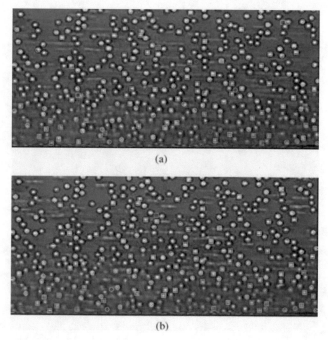

(a)

(b)

FIGURE 4.4: Tracking multiple cells from *in vitro* video microscopy. Two consecutive frames (a) and (b) are shown here.

4.5 SUMMARY

The significance of this chapter lies in the introduction of sampling methods that succeed in biomedical tracking where the Kalman filter fails. The particle filter is a flexible target-tracking tool that accommodates nonlinearity and non-Gaussianity. These models also generalize well to the multi-target case—where we wish not only to track each target but also to model the interaction between targets.

CHAPTER 5

Tracking Shapes by Sampling

"Either I mistake your shape and making quite, or else
you are that shrewd and knavish sprite."

—Shakespeare

So the LORD God said to the serpent, "Because you
have done this, cursed are you above all the livestock
and all the wild animals! You will crawl on your belly
and you will eat dust all the days of your life."

—Genesis 3:14

5.1 OVERVIEW

In automated inspection (e.g., circuit inspection), we encounter mostly a mundane assortment of static rectangular objects. Remote sensing often reveals arbitrary static shapes that can be located by straightforward correlation techniques, as with finding the distinct shape of Virginia's eastern shore. The shapes in military imaging become slightly more exciting and unpredictable, with rod-like turrets sticking out of a pile of blocks (*i.e.*, a tank). These tanks often move and sometimes shoot back. In imaging the heart, the prostate, the brain, the lungs, and the vasculature, we find moving objects of deformable but often identifiable shapes. We could, in principle, apply the point-based methods of Chapter 3 to track these objects, but in the process, we would forfeit the richness of the shape information.

In looking at these biomedical-tracking problems from a shape-based perspective, we first discover the challenge of accurate segmentation, which is treated in our companion book, *Biomedical Image Analysis: Segmentation*. A second challenge encountered in tracking shapes is the difficulty to obtain a motion model for the shape. Consider the myocardial border, for example. The myocardium behaves differently at different locations within the muscle at different points in the cardiac cycle, making it difficult to use a global motion model. A third challenge is deformation. An object such as the myocardium could certainly not be well modeled with a rigid framework.

In response to the second and third challenges, we systematically outline viable shape-based strategies for tracking biomedical objects. We start with a rigid model that is transformed in the affine or projective geometry sense, and then progress to stochastic models that allow the contour to jump around in the search for the optimal delineation. We conclude the chapter by adapting the Bayesian multi-target tracker framework to the problem of tracking shapes, producing a powerful shape-based tracking paradigm.

5.2   TRACKING RIGID SHAPES

A nuisance of the biomedical image analysis world is deformability. However, it is not rare that we encounter objects for tracking that are only slightly deformable and can be modeled via rigid body deformations, such as affine or projective transformations. Examples include tracking of bone, cartilage, the cornea, the retina, and at a larger scale, tracking of the head or limbs. The aim of this section is to provide techniques to address tracking when objects can be modeled or approximated by rigid body transformation. We classify these techniques as tracking by *affine* or *projective snakes* with respect to the underlying transformation.

5.2.1   Tracking by Affine and Projective Snakes

Often in a temporal sequence of image frames, the constituent objects change in scale. An object may gradually appear larger (or smaller) as a result of relative movement between the object and the camera. Due to the relative camera and object movement, the viewing angle also changes. For example, a circular shaped object may appear to gradually turn into an elliptic shape, which might appear more elongated as time passes (*i.e.*, an increase in eccentricity). Such change in object scale and viewpoint can often be modeled by an affine transformation or more generally, by a projective transformation on the object. Hopefully, we can give the reader appreciation of how this type of modeling can help track a shape in a noisy and cluttered environment. We first describe affine transformation in connection with snake evolution and then express the same under the context of projective transformation.

5.2.2   Affine Snakes for Tracking

For the sake of simplicity, we limit the projective and affine transformations to 2D shapes or contours. A point in 2D space is said to have undergone an affine transformation when the new transformed position (x, y) is related to the position (x', y') in the previous frame as follows:

$$\begin{bmatrix} x \\ y \end{bmatrix} = \begin{bmatrix} a & b \\ c & d \end{bmatrix} \begin{bmatrix} x' \\ y' \end{bmatrix} + \begin{bmatrix} t_x \\ t_y \end{bmatrix}, \tag{5.1}$$

where (t_x, t_y) is the translation. The matrix $\begin{bmatrix} a & b \\ c & d \end{bmatrix}$ is responsible for rotation, scaling, and shearing. If the determinant of this matrix is positive, then the transformation is known as orientation-preserving. If the determinant is negative, then the transformation is called orientation-reversing (as with a mirror reflection). These six independent parameters $(t_x, t_y, a, b, c,$ and $d)$ are called parameters of the affine transformation. When we say a 2D shape (such as contour) has been affine-transformed, we mean that every point on the shape follows the transformation rule

(5.1). Thus, for a shape, all the points in the contour must have the same associated affine parameter values.

Consider, for example, a rectangle that has been transformed into a parallelogram, or a circle into an ellipse. The advantage of affine/projective modeling in these cases is quite appealing—only a few parameters capture the transformation of an arbitrary shape, regardless of the magnitude of the displacement. For example, a snake movement can be modeled by affine transformation. We call such a snake an *affine snake*. In such a case, we do not need to track all the snaxels, we just need to keep track these six parameter values. Further, unlike a traditional snake (as discussed in Chapter 2), which is often attracted to noise or clutter, an affine snake can be designed that is much more robust to noise and clutter. On the other hand, if the object deviates substantially from rigidity, this approach is ill advised.

For the time being, let us imagine that instead of tracking, we are asked to search for an object in an image. The object model allows only affine transformations of a given contour, so that each instance is characterized by an initial contour $\{X'_i, Y'_i\}_{i=1}^{N}$ and the six parameters $(t_x, t_y, a, b, c,$ and $d)$ used to transform this contour. A brute force search method would be to exhaustively explore the affine transformation parameter space. Unless the admissible parameter values are few in cardinality, it is practically impossible to check the transformations for all possible combinations of the parameter values.

In order to locate this "best match" contour, we could minimize an external energy that depends only on the contour location as follows:

$$E(\{X_i, Y_i\}_{i=1}^{N}) = -\frac{1}{N} \sum_{i=1}^{N} f(X_i, Y_i), \qquad (5.2)$$

where $i = 1, \ldots, N$ labels the N snaxels. Once again, $f(x, y) = |\nabla I(x, y)|^2$ is an edge indicator in the image that depends on the gradient magnitude. Here, for each i, (X_i, Y_i) and (X'_i, Y'_i) are related by the affine transformation (5.1).

We want to minimize (5.2) by varying the position of the contour $\{X_i, Y_i\}_{i=1}^{N}$; however, we vary the contour position $\{X_i, Y_i\}_{i=1}^{N}$ indirectly by fixing the current snaxels $\{X'_i, Y'_i\}_{i=1}^{N}$ and only making changes to the affine parameter values in (5.1) to achieve new contour locations. Because of the affine transformation relationship (5.1), the energy functional (5.2) is a function of only the six affine parameters:

$$E(t_x, t_y, a, b, c, d) = -\frac{1}{N}\sum_{i=1}^{N} f(X_i, Y_i) = -\frac{1}{N}\sum_{i=1}^{N} f(aX'_i + bY'_i + t_x, cX'_i$$
$$+ dY'_i + t_y). \qquad (5.3)$$

Note that unlike the snake energy functional, there is no "shape" constraint in the energy functional (5.3). However, to achieve stability in object tracking, we can impose a constraint on the affine parameters. For example, we may want to limit the change in the determinant of the matrix $\begin{bmatrix} a & b \\ c & d \end{bmatrix}$ as suggested in [29]. The determinant of the affine matrix is $\left|\begin{bmatrix} a & b \\ c & d \end{bmatrix}\right| = (ad - bc)$. The modified energy functional now becomes

$$E_{\text{affine}}(t_x, t_y, a, b, c, d) = -\frac{1}{N}\sum_{i=1}^{N} f(X_i, Y_i) + \lambda(ad - bc)^2, \qquad (5.4)$$

where λ is a tunable positive number controlling the importance of the constraint in the energy functional. Penalizing the determinant of the four-element matrix limits change in scaling (object size)—for example, an increase in scaling for the x-direction must be balanced by a decrease in scaling in the y-direction.

Let us now move on to minimize the energy functional using the rules from calculus. Here we take the partial derivatives of the function with respect to the function arguments, equate the partial derivatives to zero, and solve the resulting equations. These partial derivatives are given by

$$\frac{\partial E_{\text{affine}}}{\partial t_x}(t_x, t_y, a, b, c, d) = -\frac{1}{N}\sum_{i=1}^{N} \frac{\partial f}{\partial x}(X_i, Y_i),$$

$$\frac{\partial E_{\text{affine}}}{\partial t_y}(t_x, t_y, a, b, c, d) = -\frac{1}{N}\sum_{i=1}^{N} \frac{\partial f}{\partial y}(X_i, Y_i),$$

$$\frac{\partial E_{\text{affine}}}{\partial a}(t_x, t_y, a, b, c, d) = -\frac{1}{N}\sum_{i=1}^{N} \frac{\partial f}{\partial x}(X_i, Y_i)X'_i + 2\lambda(ad - bc)d,$$

$$\frac{\partial E_{\text{affine}}}{\partial b}(t_x, t_y, a, b, c, d) = -\frac{1}{N}\sum_{i=1}^{N}\frac{\partial f}{\partial x}(X_i, Y_i)Y'_i - 2\lambda(ad - bc)c,$$

$$\frac{\partial E_{\text{affine}}}{\partial c}(t_x, t_y, a, b, c, d) = -\frac{1}{N}\sum_{i=1}^{N}\frac{\partial f}{\partial y}(X_i, Y_i)X'_i - 2\lambda(ad - bc)b,$$

$$\frac{\partial E_{\text{affine}}}{\partial d}(t_x, t_y, a, b, c, d) = -\frac{1}{N}\sum_{i=1}^{N}\frac{\partial f}{\partial y}(X_i, Y_i)Y'_i + 2\lambda(ad - bc)a.$$

The next task is to equate these equations to zero and solve for the parameter values. We can adopt the same strategy of gradient descent here as well, *i.e.*, we solve for the parameters via the following gradient descent equations:

$$\frac{\partial t_x}{\partial \tau} \propto -\frac{\partial E_{\text{affine}}}{\partial t_x}(t_x, t_y, a, b, c, d) = \frac{1}{N}\sum_{i=1}^{N}\frac{\partial f}{\partial x}(X_i, Y_i),$$

$$\frac{\partial t_y}{\partial \tau} \propto -\frac{\partial E_{\text{affine}}}{\partial t_y}(t_x, t_y, a, b, c, d) = \frac{1}{N}\sum_{i=1}^{N}\frac{\partial f}{\partial y}(X_i, Y_i),$$

$$\frac{\partial a}{\partial \tau} \propto -\frac{\partial E_{\text{affine}}}{\partial a}(t_x, t_y, a, b, c, d) = \frac{1}{N}\sum_{i=1}^{N}\frac{\partial f}{\partial x}(X_i, Y_i)X'_i - 2\lambda(ad - bc)d,$$

$$\frac{\partial b}{\partial \tau} \propto -\frac{\partial E_{\text{affine}}}{\partial b}(t_x, t_y, a, b, c, d) = \frac{1}{N}\sum_{i=1}^{N}\frac{\partial f}{\partial x}(X_i, Y_i)Y'_i + 2\lambda(ad - bc)c,$$

$$\frac{\partial c}{\partial \tau} \propto -\frac{\partial E_{\text{affine}}}{\partial c}(t_x, t_y, a, b, c, d) = \frac{1}{N}\sum_{i=1}^{N}\frac{\partial f}{\partial y}(X_i, Y_i)X'_i + 2\lambda(ad - bc)b,$$

$$\frac{\partial d}{\partial \tau} \propto -\frac{\partial E_{\text{affine}}}{\partial d}(t_x, t_y, a, b, c, d) = \frac{1}{N}\sum_{i=1}^{N}\frac{\partial f}{\partial y}(X_i, Y_i)Y'_i - 2\lambda(ad - bc)a,$$

where as before τ is the (pseudo) time used to denote time evolution of the differential update equations.

As before with snake energy minimization, one needs to use a small time step ζ while updating snaxel locations in each iteration. However, this update could often be numerically unstable unless the time step size ζ is very small. To circumvent this problem we can apply the Langevin–Hastings principle in the Markov chain Monte Carlo (MCMC) method [30]. Recall the Metropolis–Hastings (MH) algorithm, where in each iteration of sampling, we need to have a new sample for the

Markov chain, then we compute the MH ratio, and accept or reject the new sample accordingly. The following steps outline the $(\tau + 1)$th iteration for updating the transformation parameters:

Step 1: Given the parameter values at time τ, for a selected time step ζ, compute the following:

$$t_x = t_x^\tau - \zeta \frac{\partial E_{\text{affine}}}{\partial t_x^\tau} + \sqrt{2\zeta}\, w_1^{\tau+1}, \qquad t_y = t_y^\tau - \zeta \frac{\partial E_{\text{affine}}}{\partial t_y^\tau} + \sqrt{2\zeta}\, w_2^{\tau+1},$$

$$a = a^\tau - \zeta \frac{\partial E_{\text{affine}}}{\partial a} + \sqrt{2\zeta}\, w_3^{\tau+1}, \qquad b = b^\tau - \zeta \frac{\partial E_{\text{affine}}}{\partial b} + \sqrt{2\zeta}\, w_4^{\tau+1},$$

$$c = c^\tau - \zeta \frac{\partial E_{\text{affine}}}{\partial c} + \sqrt{2\zeta}\, w_5^{\tau+1}, \qquad d = d^\tau - \zeta \frac{\partial E_{\text{affine}}}{\partial d} + \sqrt{2\zeta}\, w_6^{\tau+1}.$$

Step 2: Compute MH ratio:

$$R = \frac{\exp[-E_{\text{affine}}(t_x, t_y, a, b, c, d)]}{\exp[-E_{\text{affine}}(t_x^\tau, t_y^\tau, a^\tau, b^\tau, c^\tau, d^\tau)]}.$$

Step 3: Generate a uniformly distributed random number $u \in [0, 1]$

If $R \geq u$ assign $t_x^{\tau+1} \leftarrow t_x$, $t_y^{\tau+1} \leftarrow t_y$, $a^{\tau+1} \leftarrow a$, $b^{\tau+1} \leftarrow b$, $c^{\tau+1} \leftarrow c$, $d^{\tau+1} \leftarrow d$,

Else assign

$t_x^{\tau+1} \leftarrow t_x^\tau$, $t_y^{\tau+1} \leftarrow t_y^\tau$, $a^{\tau+1} \leftarrow a^\tau$, $b^{\tau+1} \leftarrow b^\tau$, $c^{\tau+1} \leftarrow c^\tau$, $d^{\tau+1} \leftarrow d^\tau$.

The ws in Step 1 of this algorithm are independent samples from a Gaussian distribution with zero-mean and a variance of one. A problem with the MH algorithm is the diagnosis of convergence. While there are several methods available to evaluate convergence [31], often in practice, we obtain empirical convergence knowledge from several experiments with the data at hand. Once sufficiently many samples are generated through the Langevin–Hastings algorithm, we compute their mean to determine the affine parameters and consequently the final contour position. Thus, the mean of all samples for parameter a is used as the new a, for example. The Langevin–Hastings algorithm is quite versatile and can also be applied to general snake evolutions.

There is yet another way to minimize the affine snake functional [32]. In this case, we typically do not utilize the constraint on the affine parameters [the $(ad - bc)^2$ term], so we describe the curve evolution via energy functional (5.2). Let us assume that at iteration number τ, the position of the curve is $\{X_i^\tau, Y_i^\tau\}_{i=1}^N$. The algorithm first computes the unconstrained displacement of the contour $\{X_i', Y_i'\}_{i=1}^N$ by gradient descent. Next, the affine parameters are computed according to the least squares criterion. To express the least squares formulation, we first make a discrete description of the continuous contour by N contour points or snaxels; so the contours are represented now as (X_i^τ, Y_i^τ) and (X_i', Y_i') for $i = 1 - N$. The affine parameters can be found by solving the over-determined system of linear equations by least squares methods:

$$a X_i^\tau + b Y_i^\tau + t_x = X_i',$$
$$c X_i^\tau + d Y_i^\tau + t_y = Y_i', \text{ for } i = 1, \ldots, N.$$

Note that the number of equations is $2N$, whereas the number of variables is only 6. Since typically N is greater than 3, the system is over-determined. As a numerical method, this method is superior to the gradient descent method, because in the gradient descent approach, the time step plays a crucial role in stability.

If the least squares solution is $(t_x^*, t_y^*, a^*, b^*, c^*, d^*)$, then the contour movement occurs via $X_i^{\tau+1} \leftarrow (a^* X_i^\tau + b^* Y_i^\tau + t_x^*)$ and $Y_i^{\tau+1} \leftarrow (c^* X_i^\tau + d^* Y_i^\tau + t_y^*)$, for $i = 1 - N$. The first iteration starts with the following assignments: $X_i^1 \leftarrow X_{0,i}$ and $Y_i^1 \leftarrow Y_{0,i}$, for $i = 1 - N$, where $(X_{0,i}, Y_{0,i})$ is the discrete version of the initial contour $(X_0(s), Y_0(s))$ mentioned before. The algorithm halts when the solution corresponds to a local minimum in the energy.

So, in an actual biomedical tracking application, how do we use the affine contours? One straightforward way to combine tracking with object delineation is by way of the nearest neighborhood assumption. In this case, as earlier, we take the affine

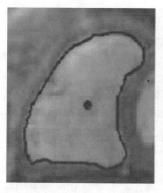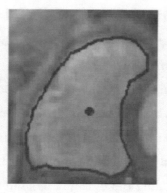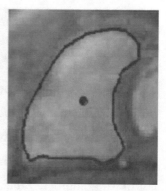

FIGURE 5.1: Three consecutive frames of the right ventricle (RV) from a first pass magnetic resonance study. An affine snake is used to delineate the RV in each frame. The centroid of the snake is denoted by a dot in the RV.

snake from the previous frame and place it on the current frame, and delineate the object by minimizing (5.4).

Figure 5.1 shows three consecutive frames of the right ventricle from a first pass magnetic resonance study. An affine snake (realized by gradient descent on (5.4)) is used to delineate the right ventricle in each frame.

5.2.3 Projective Snakes for Tracking

The projective transformation is a more general transformation. Unlike the affine transformation, a projective transformation can account for foreshortening of an object within a temporal sequence of image frames. We say (x, y) is a projective transformation of (x_0, y_0) when they are related by

$$x = \frac{ax_0 + by_0 + c}{gx_0 + hy_0 + 1}, \qquad y = \frac{dx_0 + ey_0 + f}{gx_0 + hy_0 + 1},$$

where $a, b, \ldots, g, h$ are the eight independent parameters of the transformation. A projective snake can now be defined in the same way as in the affine case. To use a projective model in lieu of the affine one to implement tracking, we can apply the projective model in the same energy minimization process discussed above for the affine case.

5.3   TRACKING DEFORMABLE SHAPES

5.3.1   Stochastic Snakes for Tracking: Simulated Annealing and Deterministic Annealing

The snakes or active contours discussed thus far in the book, aside from the dynamic programming models, have been implemented by first forming an energy functional that describes snake quality and then by performing gradient descent on the energy. As with any combinatorial optimization problem, there are several methods of computing solutions, including the dynamic programming solution discussed in Chapter 2. In this section, we consider alternate implementations based on methods that can "escape" local minima in the energy.

But before we leave gradient descent for more sophisticated remedies, it must be mentioned that gradient descent (and associated techniques such as conjugate gradient approaches and other greedy methods) may give the most attractive solution in certain applications. First, the gradient descent solution is generally expedient. Second, the descent solution could avoid target-hopping in a tracking application. In an application where there exist several similar targets in a congested area (e.g., cell tracking), the optimal solution in terms of energy could be the contour of a neighboring target, instead of the selected target. So, in considering global optimality, one must be careful to avoid destroying target correspondence.

However, in situations where noise dominates and the target signal is weak, as in many biomedical tracking applications, an approach may be needed to optimize a nonconvex energy functional. In this case, the nonconvexity usually arises from conflicting constraints—such as the conflict between generating a smooth contour and locating the contour at the points of highest gradient magnitude. The phenomenon of getting stuck in a local minimum may manifest itself as the snake resting on a random conglomeration of noise or image clutter, instead of delineating the intended tissue, organ, cell, *etc.*

5.3.1.1   Simulated Annealing

To combat the failure to escape local minima in the energy, methods based on stochastic simulated annealing have been probed [33, 34, 35]. Consider a one-dimensional energy function with energy as the y-axis and the different so-lutions arranged at the x-axis. A gradient descent technique always makes moves downhill in energy toward the closest valley—a local minimum (which indeed *could* be the global minimum). The stochastic simulated annealing method, on the other hand, allows the solution to "hill-climb" at high temperatures. Here, an artificial parameter, called the *temperature*, is used to describe the possibility of jumping to higher energy states. The idea is that the algorithm can start at a high temperature and allow moves to higher energy states in order to pop out of local minima before the solution is "cooled" allowing refinement into a deeper local minimum. According to the theory developed by Geman and Geman, the simu-lated annealing algorithm can find the global minimum given a sufficient number of steps [36]. Unfortunately, this number of moves explodes as $O(K^N)$ (at each annealing temperature), where N is the number of variables (snake samples or snaxels in this application), and K is the number of possible positions for each snaxel.

Imagine an interconnected state diagram that represents the entire set of possible solutions for a snake (represented by the possible positions of the snake samples or snaxels). The interconnections represent possible one-snaxel changes in position; the connected states are neighbors in the solution space. Now, consider stochastic jumps in this solution space. Each jump would have a probability of being considered and then a probability of being enacted once considered. In order to explore the solution space and not become stuck in a locally optimal solution or set of solutions, "risky" jumps to higher energy (lower quality) states would be considered early in the optimization process. Such foolish jumps would decrease in probability as the optimization process continued. *This stochastic exploration is the essence of the stochastic simulated annealing (SA) solution.*

In SA, the solution space Ω is represented by a Markov chain. In this case, consider one snake $\mathbf{C}$, which corresponds to one state in the Markov chain. Note that a given snake is defined by the fixed position of the N discrete snaxels. For a Markovian $\mathbf{C}$, it is required that the probability $P(\mathbf{C} = \mathbf{C}_j) > 0, \forall C_j \in \Omega$, and that new solutions are generated only within a neighborhood $\aleph(\mathbf{C}_j)$ of solution $\mathbf{C}_j$.

Then the Markov chain representing the solutions for $\mathbf{C}$ can be modeled by the Gibbs distribution [36]

$$P(\mathbf{C} = \mathbf{C}_j) = \frac{1}{Z} e^{-E(\mathbf{C}_j)/T} \tag{5.5}$$

where the partition function is given by

$$Z = \sum_{\forall \mathbf{C}_i \in \Omega} e^{-E(\mathbf{C}_i)/T}, \tag{5.6}$$

where T is the annealing temperature and E is the energy functional that quantifies the solution quality (lower E is preferable in our convention). At high values of T, the SA Markov chain has a uniform distribution in which all solutions are equally likely. If annealed properly [37] according to a slow logarithmic temperature decrease $T(t) \geq T_0 / \log(1 + t)$ for iteration t, the SA Markov chain will converge to a uniform distribution over the global minima in the energy functional E. Typically, a "fast" geometric schedule [37] is employed in practical implementations where $T(t) = T(t-1)\tau$ and τ is a reduction factor slightly less than 1 (e.g., 0.9).

As we mentioned, simulating annealing keeps track of the current solution by way of a Markov chain. Here, we move between solutions (states in the Markov chain) by updating the snake sample positions (snaxel positions). The moves are first *generated* and then *accepted* on the basis of the energy change and temperature, where moves to higher energy solutions are rejected with higher probability as the temperature is lowered.

Let $\mathbf{C}_1$ and $\mathbf{C}_2$ denote tensors holding the vector of position vectors for the snake samples. The *generation function* gives the probability of generating a move from one solution $\mathbf{C}_1$ to another neighboring solution $\mathbf{C}_2$. Probability density functions such as the uniform density (over a limited neighborhood) and the Gaussian can be employed to generate new snaxel positions.

Given that the problem is Markov—that is, the probability of given snake solution depends only on the neighborhood of solutions surrounding that solution— we can derive an acceptance function using the Gibbs distribution. Let us say that solution $\mathbf{C}_2$ is generated from "current" solution $\mathbf{C}_1$. Then, this solution is accepted if [37] a uniformly distributed random variable with range [0, 1] is less than or equal to

$$A(\mathbf{C}_1, \mathbf{C}_2, T) = \frac{1}{1 + e^{[E(\mathbf{C}_2) - E(\mathbf{C}_1)]/T}},\qquad(5.7)$$

which is the acceptance function at temperature T. Else, the move is rejected.

5.3.1.2 Generalized Deterministic Annealing

SA provides a theoretically rigorous method to explore the solution space. Unfortunately, the high cost of implementation eclipses the attractiveness of the ability to avoid local minima. In looking at generalized deterministic annealing (GDA) [38] for a snake implementation to be used in biomedical tracking applications, we are seeking an implementation that is both low cost and high quality in terms of the ability to escape local minima in the snake energy functional [39]. In this section, GDA is used to compute the positions of the N discrete samples (snaxel positions) in the snake contour. Instead of making random moves to sample the solution space, GDA makes deterministic updates that estimate the stationary distribution over the solution of SA at each annealing temperature.

Essentially, the idea of GDA boils down to localizing the problem and then direct computing Markov chain stationary distributions by iteratively using the Markov chain transition probabilities. Since we cannot compute the stationary distribution directly for the complete solution space, we, in the case of snake computation, compute the stationary distribution for a Markov chain representing one of the snaxel (snake pixel) positions while keeping the remaining snaxels fixed.

To compute a snake solution for tracking via GDA, we first subdivide the solution space into "local" Markov chains that represent local variables (the K possible positions of each snake sample). Then, we approximate the stationary distribution of the local Markov chain at each annealing temperature using the SA transition probabilities. This update is an iterative process that converges to the SA stationary distribution. While updating the stationary distribution of a local

Markov chain, we utilize the mean field approximation (the expected value) for the state of neighboring local Markov chains.

The probabilistic model given by SA is employed to compute $P(\mathbf{C}_1, \mathbf{C}_2, T)$, the probability of moving from solution $\mathbf{C}_1$ to solution $\mathbf{C}_2$ at temperature T. From these transition probabilities, we can compute the expected state for each local Markov chain (and the associated distribution). SA gives us the generation function $G(\mathbf{C}_1, \mathbf{C}_2)$ and the acceptance function $A(\mathbf{C}_1, \mathbf{C}_2, T)$. The transition probabilities are simply the product of these two sequential functions:

$$P(\mathbf{C}_1, \mathbf{C}_2, T) = G(\mathbf{C}_1, \mathbf{C}_2)A(\mathbf{C}_1, \mathbf{C}_2, T), \quad \forall \mathbf{C}_2 \neq \mathbf{C}_1 \qquad (5.8)$$

Or, in the case of a self-transition where the solution remains the same, we have

$$P(\mathbf{C}_1, \mathbf{C}_1, T) = 1 - \sum_{\forall \mathbf{C}_2 \neq \mathbf{C}_1} P(\mathbf{C}_1, \mathbf{C}_2, T). \qquad (5.9)$$

Note that it is assumed that (1) all solutions have non-zero probability, (2) any solution is reachable (can be reached by a finite series of moves in the Markov chain), and (3) the generation function is symmetric:

$$G(\mathbf{C}_1, \mathbf{C}_2) = G(\mathbf{C}_2, \mathbf{C}_1).$$

Given (5.8) and (5.9), why not compute the stationary distribution of the entire Markov chain used in SA? Unfortunately, this approach would be as expensive as an exhaustive search itself. As an alternative, we localize the problem. For one snake, we create N local Markov chains representing the positions of the N samples of the discretized contour, each of which has K possible solutions. Consider a fixed center (x, y); the sample positions in the snake can be denoted by the (r, θ) position with respect to the center (x, y). (Note that the center does not change while evolving the GDA snake in the implementation given here and that θ is given such that the positive x-axis direction corresponds with $\theta = 0$.) Each of the N snake samples can be parameterized by the angle θ. So, the local Markov chain tracks the distance $C(\theta) = r$ between the curve and the initial center position in the direction θ, which has K possible solutions.

Instead of the radial parameterization, a local solution space containing a finite number of Cartesian pairs is also possible (and has been implemented) for the GDA snake. The attraction of the radial (r, θ) model lies in the simplicity of the energy terms (as only one variable $C(\theta)$ for each snaxel is computed), in the avoidance of reparameterization due to bunching/spreading of adjacent contour samples, and in the match for the application of delineating cell boundaries. The radial model also avoids exploring "illegal" solutions such as self-intersections and loops. A limitation of the radial model is that it assumes "star-shaped" boundaries.

In more global optimization probabilities, such as the traveling salesman problem, another difficulty encountered with SA is that the computation of changes in the energy functional is computationally expensive. In the snake-based tracking application discussed here, however, only the change in the local energy functional, $\Delta E_\theta(r_1, r_2)$, is required. The term $\Delta E_\theta(r_1, r_2)$ denotes the energy change associated with changing the radius at angle θ from r_1 to r_2. Actual examples of $\Delta E_\theta(r_1, r_2)$ are provided in this chapter for the cell tracking application.

In GDA, one needs to estimate the stationary distribution π. One element, $\pi_\theta(r_1)$, represents the probability of the snake sample at angle θ having a distance of r_1 from the contour center. The probabilities (simulating one stochastic move via SA) can be updated using the transition probabilities (5.8) and (5.9). In order to compute the transition probabilities, we need to design appropriate generation functions (that describe the probability of given transitions) and acceptance functions (that describe the probability of accepting such a move at temperature T).

For the GDA snake generation function, we assume that the possible transitions to new positions for a given snaxel are uniformly distributed among the K possible positions. (This decision is arbitrary; other distributions such as a Gaussian centered at the expected position, are possible.) For the acceptance function, we adopt the sigmoidal SA acceptance function of (5.7), resulting in

$$A_\theta(r_1, r_2, T) = \frac{1}{1 + e^{[E_\theta(r_2) - E_\theta(r_1)]/T}} = \frac{1}{1 + e^{\Delta E_\theta(r_1, r_2)/T}} \tag{5.10}$$

for each local Markov chain. This function gives the probability of acceptance of a move from radial position r_1 to r_2 at angle θ and temperature T.

The $(t + 1)$st update for state r_2 at the local Markov chain corresponding to the snaxel position for angle θ is computed recursively as follows:

$$\pi_\theta^{t+1}(r_2) \leftarrow \sum_{\forall r_1 \neq r_2, r_1 \in R} P_\theta(r_1, r_2, T) \pi_\theta^t(r_1) + \pi_\theta^t(r_2) \left[1 - \sum_{\forall r_3 \neq r_2, r_3 \in R} P_\theta(r_2, r_3, T) \right] \tag{5.11}$$

where t denotes the (t)th iterative update, and R denotes the K-member set of possible radial positions at sample θ. In (5.11), $P_\theta(r_1, r_2, T)$ represents the transition probability within the local Markov chain at sample θ for a transition from r_1 to r_2 at temperature T. The transition probabilities are built via the product of the generation function and the acceptance function. Here, we can exploit a special case that assumes a uniform generation (where every neighboring solution is equally likely).

Using (5.8) and (5.9), assuming a uniform generation function and the sigmoidal acceptance function of (5.10), where $A_\theta(r_1, r_2, T) = 1 - A_\theta(r_2, r_1, T)$, we have a simplified update in

$$\pi_\theta^{t+1}(r_2) \leftarrow \frac{1}{K} \sum_{\forall r_1 \in R} A_\theta(r_1, r_2, T)[\pi_\theta^t(r_1) + \pi_\theta^t(r_2)], \tag{5.12}$$

which is termed the GDA update function.

One may notice from (5.12) that the acceptance function, and in turn, the local energy change, are the only computations needed to accomplish this distribution update. However, the energy change depends on the expected state of neighboring snaxel positions. GDA utilizes a mean field approximation, which is simply the expected value for the snaxel position given the current estimate of the stationary distribution. The mean field position is computed by taking a weighted average (weighted by π_θ^t) of the K possible positions corresponding to the K local Markov chain states. So, in the case of implementing a snake, we have snaxels that can be located at K possible positions. Given the associated probabilities, we can compute a mean field position by basic weighted averaging.

5.3.1.3 Implementation of the GDA Snake for Tracking Annealing Schedule

In order to provide the ability to escape locally optimal solutions, annealing is employed for the GDA snake. The annealing schedule is simply the sequence of temperatures at which the stationary distributions of the local Markov chains are computed. Given a geometric rate of temperature decrease (say, 90% per temperature change), the main variables in the annealing schedule are the initial and final temperatures.

The acceptance function of (5.10) can be exploited to determine these initial and final temperatures. If an initial temperature is chosen that is too low, the algorithm will not be able to escape local minima and will behave in the manner of gradient descent. So, a reasonable percentage of these "bad" (positive energy change) moves should be accepted. Likewise, if the final temperature is too high, the GDA process may be halted before it converges to a local minimum in the energy. So, the probability of an increase in the energy should be very low at the final temperature.

5.3.1.4 An Example Energy Functional and Computation of Local Energy Differences

GDA is essentially an optimization technique for combinatorial optimization problems that can be localized. To apply GDA to tracking biomedical objects, a snake and the accompanying energy functional are prescribed. Here, the snake should be smooth and of a certain size and shape; it should also correspond to positions of high gradient magnitude in the image. The resulting energy functional is similar to the functional used in [10] (in [10], the snake is implemented by gradient descent), with a few exceptions that are noted.

The energy functional employed here has five additive terms. The first three energy terms (tension, rigidity, and external energy) are the typical energy terms used in any parametric snake such as in the classical Kass, Witkin, and Terzopoulos model [1]. The fourth and fifth terms enforce constraints on the snake shape and size, respectively.

Recall that the snake has a fixed center. With respect to the fixed center, $C(\theta)$ gives the distance of the snake from the center at angle θ. In this case, we sample the curve at N points separated by $2\pi/N$ radians. The tension energy term is computed via

$$E^{\text{tension}} = \sum_{\theta=\theta_0}^{\theta_{N-1}} \left\{ C\left[\left(\theta + \frac{2\pi}{N}\right)_{\text{mod } 2\pi}\right] - C(\theta) \right\}^2, \qquad (5.13)$$

where the summation is computed for the N snaxels of the snake in angular steps of $2\pi/N$. In (5.13), for the snaxel located $C(\theta)$ away from the center of the curve at angle θ, $C[(\theta + (2\pi/N)_{\text{mod } 2\pi})]$ represents the neighboring snaxel position in the counter-clockwise direction. For GDA, the critical expression is the energy difference equation needed for computation of the acceptance function, not the energy itself. The energy difference term for the tension portion of the energy is

$$\Delta E_\theta^{\text{tension}}(r_1, r_2) =$$
$$\left[\begin{array}{l} \left\{ r_2 - C\left[\left(\theta - \frac{2\pi}{N}\right)_{\text{mod } 2\pi}\right] \right\}^2 + \left\{ C\left[\left(\theta + \frac{2\pi}{N}\right)_{\text{mod } 2\pi}\right] - r_2 \right\}^2 \\ - \left\{ r_1 - C\left[\left(\theta - \frac{2\pi}{N}\right)_{\text{mod } 2\pi}\right] \right\}^2 - \left\{ C\left[\left(\theta + \frac{2\pi}{N}\right)_{\text{mod } 2\pi}\right] - r_1 \right\}^2 \end{array} \right]. \qquad (5.14)$$

Hence, (5.14) gives the change in tension energy for changing the snake position at angle θ from r_1 to r_2. The total energy change is simply the sum of the energy changes associated with each of the five terms. In (5.14), the term $C[(\theta - (2\pi/N))_{\text{mod } 2\pi}]$ represents the neighboring snaxel position (to $C(\theta)$) in the clockwise direction.

The second snake smoothness term, called the *rigidity* term, can be expressed by

$$E^{\text{rigid}} = \sum_{\theta=\theta_0}^{\theta_{N-1}} \left\{ C\left[\left(\theta - \frac{2\pi}{N}\right)_{\text{mod } 2\pi}\right] - 2C(\theta) + C\left[\left(\theta + \frac{2\pi}{N}\right)_{\text{mod } 2\pi}\right] \right\}^2.$$

(5.15)

Then, the energy difference term used in the GDA update equation is

$$\Delta E_\theta^{\text{rigid}}(r_1, r_2) = \begin{bmatrix} \left\{ C\left[\left(\theta - \frac{4\pi}{N}\right)_{\text{mod } 2\pi}\right] - 2C\left[\left(\theta - \frac{2\pi}{N}\right)_{\text{mod } 2\pi}\right] + r_2 \right\}^2 \\ + \left\{ C\left[\left(\theta - \frac{2\pi}{N}\right)_{\text{mod } 2\pi}\right] - 2r_2 + C\left[\left(\theta + \frac{2\pi}{N}\right)_{\text{mod } 2\pi}\right] \right\}^2 \\ + \left\{ r_2 - 2C\left[\left(\theta + \frac{2\pi}{N}\right)_{\text{mod } 2\pi}\right] + C\left[\left(\theta + \frac{4\pi}{N}\right)_{\text{mod } 2\pi}\right] \right\}^2 \\ - \left\{ C\left[\left(\theta - \frac{4\pi}{N}\right)_{\text{mod } 2\pi}\right] - 2C\left[\left(\theta - \frac{2\pi}{N}\right)_{\text{mod } 2\pi}\right] + r_1 \right\}^2 \\ - \left\{ C\left[\left(\theta - \frac{2\pi}{N}\right)_{\text{mod } 2\pi}\right] - 2r_1 + C\left[\left(\theta + \frac{2\pi}{N}\right)_{\text{mod } 2\pi}\right] \right\}^2 \\ - \left\{ r_1 - 2C\left[\left(\theta + \frac{2\pi}{N}\right)_{\text{mod } 2\pi}\right] + C\left[\left(\theta + \frac{4\pi}{N}\right)_{\text{mod } 2\pi}\right] \right\}^2 \end{bmatrix}.$$

(5.16)

Typically, the external force for a snake is realized by maximizing the contour integral of the gradient magnitude for the contour specified by the snake. If $I[C(\theta), \theta]$ is the image intensity at the polar (snake) position $[C(\theta), \theta]$, then

$$E^{\text{ext}} = -\sum_{\theta=\theta_0}^{\theta_{N-1}} |\nabla I[C(\theta), \theta]|,$$

(5.17)

and, the corresponding energy difference needed for (update) is

$$\Delta E_\theta^{\text{ext}}(r_1, r_2) = [|\nabla I(r_1, \theta)| - |\nabla I(r_2, \theta)|].$$

(5.18)

It is worth noting here that the difficulty in locating a distant object using a gradient descent approach is caused by the dependence on the local image gradient as in (5.17). Unless the initial position of the snake is very close (within a few pixels) to the desired boundary, the snake computed by gradient descent will not be attracted to the boundary. Due to this dilemma, modifications to (5.17) that extend the influence of the boundaries have been explored as discussed in Chapter 2 of this book (*viz.* gradient vector flow and motion gradient vector flow). The GDA and SA approaches to computing the snake position avoid smoothing of the gradient force by exploiting an exploration of the solution space that is not limited to the immediate neighborhood of the initial snaxel positions.

For the application of tracking cells, additional energy constraints tailored to the approximately known size and shape of the cells are required. The shape constraint, as introduced in [10] enforces a penalty for any radial distance that deviates from the average radius, which is sufficient when using gradient descent. But, for SA and GDA, where radical moves are possible at high temperatures, such a formulation may lead to illegal configurations such as a "C"-shaped contour that has doubled over itself. To remedy this problem, the shape constraint is modified as follows:

$$E^{\text{shape}} = \sum_{\theta=\theta_0}^{\theta_{N-1}} \{C(\theta) + C[(\theta + \pi)\,_{\text{mod}\,2\pi}] - 2\bar{r}\}^2 \tag{5.19}$$

where $C(\theta) + C[(\theta + \pi)\,_{\text{mod}\,2\pi}]$ measures the diameter w.r.t. a given snake sample at θ, and $\bar{r}$ is the average radial distance from the snake center computed by averaging $C(\theta)$ over all θ. The energy difference term used in SA and GDA is then

$$\begin{aligned}
\Delta E_\theta^{\text{shape}}&(r_1, r_2) \\
&= [(r_2 + C[(\theta + \pi)\,_{\text{mod}\,2\pi}] - 2\bar{r})^2 - (r_1 + C[(\theta + \pi)\,_{\text{mod}\,2\pi}] - 2\bar{r})^2] \\
&= r_2^2 - r_1^2 + C[(\theta + \pi)\,_{\text{mod}\,2\pi}](r_2 - r_1) - 2\bar{r}(r_2 - r_1)
\end{aligned} \tag{5.20}$$

For many biomedical applications, such as the cell-tracking application, the approximate size of the desired object is known *a priori*. In such circumstances, given an expected radius ρ, a size constraint can be implemented as follows:

$$E^{\text{size}} = \sum_{\theta=\theta_0}^{\theta_{N-1}} [C(\theta) - \rho]^2. \tag{5.21}$$

The energy difference for changing a snake position from r_1 to r_2 is

$$\Delta E_\theta^{\text{size}}(r_1, r_2) = [(r_2 - \rho)^2 - (r_1 - \rho)^2]. \tag{5.22}$$

To summarize, for the cell detection and tracking application, the GDA snake algorithm can be described by the following steps. First, we determine the initial and final annealing temperatures so that a high percentage of positive energy changes will be made initially and that virtually no positive energy changes will occur at the minimum temperature. For the first frame in a tracking sequence, we initialize the center position (x, y) manually (or by means of some other automated detection scheme). After the first frame, the initial center position can be determined by the center of mass of the final snake computed for the previous frame.

In this example, we assume N snaxels where each has K possible positions. With GDA, we are tracking the stationary distributions of the Markov chains that represent the solution space for each snaxel. Each of the K-length stationary distribution estimates for each of the N contour samples are initialized at the *trivial state*, where $\pi_\theta^0(\cdot) = 1/K$. For each of the N contour samples, we compute the K^2 possible energy changes (since each of the K possible positions could be changed to any of the other K possible positions) using a weighted sum of specified energy changes. Note that only $(1/2)K^2$ energy changes need to be computed in reality, since $\Delta E_\theta(r_1, r_2) = -\Delta E_\theta(r_2, r_1)$. Using the energy changes, we then compute the K elements of the stationary distribution estimate for each of the N samples using (5.12). After K sweeps are made, we then reduce the temperature geometrically until the final temperature is reached.

To compare the gradient descent solution of Chapter 2, the SA solution and the GDA solution from this chapter, cell tracking examples are shown in Figs. 5.2 and 5.3. In the first example (Fig. 5.2), all three snake solutions are able to track the cell border. In the second example (Fig. 5.3), however, the gradient descent solution becomes stuck in a local minimum of energy that corresponds to clutter, not the intended cell boundary. One may observe that tracking is slightly more difficult than segmentation itself, as tracking is segmentation of a target in motion.

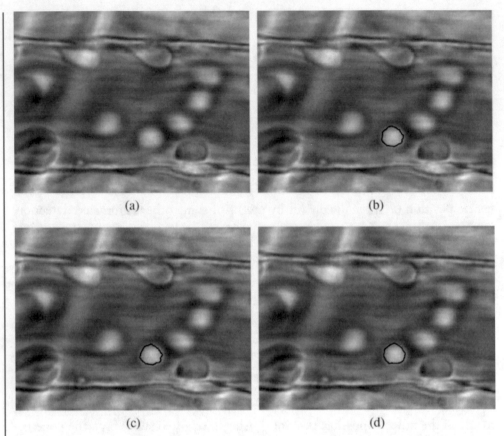

FIGURE 5.2: (a) Original video microscopy frame of a leukocyte rolling along the endothelium; (b) successful capture of the moving leukocyte by gradient descent; (c) success via SA; (d) success via GDA.

5.3.2　Sequential Bayesian Formulation

In this section, we find another probabilistic approach to shape-based tracking, one that has innumerable applications in the biomedical world and elsewhere. The aim of this section is to describe the sequential Bayesian formulation of tracking for shapes represented by contours. The interesting message to the readers at this point is that we have already developed this formulation in connection with multi-target tracking (MTT) in Section 4.2. *How so?* A closed or an open contour is nothing but a collection of points similar to a collection of point-like (shapeless) targets as in the MTT scenario. The only difference is that the points on a contour are ordered.

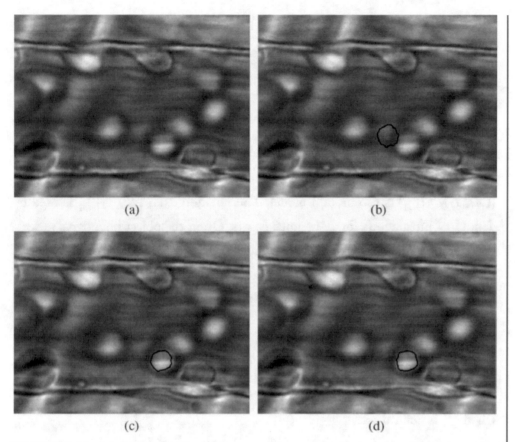

FIGURE 5.3: (a) Original video microscopy frame of a leukocyte rolling along the endothelium—taken 1/3 s after the frame in Fig. 5.2(a); (b) unsuccessful capture of the moving leukocyte by gradient descent; (c) successful capture via SA; (d) success via GDA.

Hence, the sequential Bayesian MTT formulation for the unified tracking (UT) method should serve as the computational framework here. However, we need to maintain the ordering of the points here.

Let $\{(X_i, Y_i)_t\}_{i=1}^N$ denote N contour points or snaxels that describe a snake in a discrete setting at time point t. We can simply replace the recursive equations of UT by the following:

$$
\begin{aligned}
&p[\{(X_i, Y_i)_t\}_{i=1}^N | Z_{1:t}] \\
&= \frac{p(Z_t | \{(X_i, Y_i)_t\}_{i=1}^N) p(\{(X_i, Y_i)_t\}_{i=1}^N | Z_{1:t-1})}{\int p(Z_t | \{(X_i, Y_i)_t\}_{i=1}^N) p(\{(X_i, Y_i)_t\}_{i=1}^N | Z_{1:t-1}) d\{(X_i, Y_i)_t\}_{i=1}^N}, \quad (5.23)
\end{aligned}
$$

where $p[\{(X_i, Y_i)_t\}_{i=1}^{N} | Z_{1:t-1}]$ is a joint prior density for the entire snake and is defined as follows:

$$p[\{(X_i, Y_i)_t\}_{i=1}^{N} | Z_{1:t-1}]$$
$$= \int p(\{(X_i, Y_i)_t\}_{i=1}^{N} | \{(X_i, Y_i)_{t-1}\}_{i=1}^{N}) p(\{(X_i, Y_i)_{t-1}\}_{i=1}^{N} | Z_{1:t-1}) d\{(X_i, Y_i)_{t-1}\}_{i=1}^{N}.$$

(5.24)

Here, $p(Z_t | \{(X_i, Y_i)_t\}_{i=1}^{N})$ is the likelihood of the snake, and $p(\{(X_i, Y_i)_t\}_{i=1}^{N} | \{(X_i, Y_i)_{t-1}\}_{i=1}^{N})$ is the motion model. The notation $d\{(X_i, Y_i)_{t-1}\}_{i=1}^{N}$ is used to denote N integrations for N contour points. Often the object boundary we seek is somewhat smooth, so the snaxels should make a smooth contour. *How is this smoothness imposed in the aforementioned sequential Bayesian formulation?* The key is the so-called motion model probability, which we can write as

$$p[\{(X_i, Y_i)_t\}_{i=1}^{N} | \{(X_i, Y_i)_{t-1}\}_{i=1}^{N}] \propto \prod_{i=1}^{N} f_1 [(X_i, Y_i)_t, (X_{i+1}, Y_{i+1})_t]$$
$$\times \prod_{i=1}^{N} f_2 [(X_i, Y_i)_t, (X_i, Y_i)_{t-1}],$$

where f_1 is defined as

$$f_1[(X_i, Y_i)_t, (X_{i+1}, Y_{i+1})_t] = \exp \left\{ -\frac{d^2[(X_i, Y_i)_t, (X_{i+1}, Y_{i+1})_t]}{2\sigma_s^2} \right\},$$

with d denoting the Euclidean distance between two points, and σ_s representing the standard deviation of this Gaussian density. The function f_2 can be any probability density function $p[(X_i, Y_i)_t | (X_i, Y_i)_{t-1}]$ describing the motion for the ith snaxel. The function f_1 imposes smoothness on the contour. The motion model probability f_2 actually defines a Markov Random Field (MRF) on the contour. Interested readers are referred to [40] for a nice introduction to MRFs in the image-processing context.

Once we have a likelihood density, we can generate samples for the snaxels via MCMC methods as described in Section 4.3 and estimate the snake or the contour locations from the samples via the mode or mean. We now provide a case study

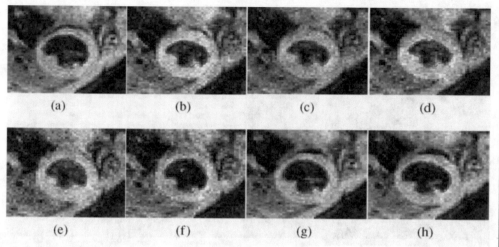

FIGURE 5.4: Mouse heart tracking: the endocardial border in the left ventricle is tracked in eight frames spanning the cardiac cycle.

based on the aforementioned formulation. The application is myocardial border tracking in MR mouse heart image sequences.

Figure 5.4 illustrates such an image showing the myocardial border in mouse heart. To apply the sequential MCMC algorithm we need to define (1) initial snaxel distribution, $p(\{(X_i, Y_i)_0\}_{i=1}^N)$, (2) the probability density $p[(X_i, Y_i)_t|(X_i, Y_i)_{t-1}]$ to fully describe the "motion model", and (3) the likelihood density $p(Z_t|\{(X_i, Y_i)_t\}_{i=1}^N)$. We have chosen the likelihood as the density associated with the GICOV statistic (see Section 4.4). Often for simplicity or for the lack of an available joint motion model, one prefers to assume that the motion model factors over the snaxels. We assume that initial contour delineation provides the initial snaxel distribution. The motion model $p[(X_i, Y_i)_t|(X_i, Y_i)_{t-1}]$ is hard to obtain. However the availability of the tagged imagery (see Fig. 5.5) can lead to a probability model for motion. With tagged images, we essentially have motion vectors for each image pixel. Let $f[(X_i, Y_i)_{t-1}]$ denote the trajectory of the point $[(X_i, Y_i)_{t-1}]$ on the tagged image. In the discrete sense, we may say that $f[(X_i, Y_i)_{t-1}]$ is a finite set of points that can be easily obtained from the tagged image given the point coordinates $[(X_i, Y_i)_{t-1}]$. We make the set $f[(X_i, Y_i)_{t-1}]$ of

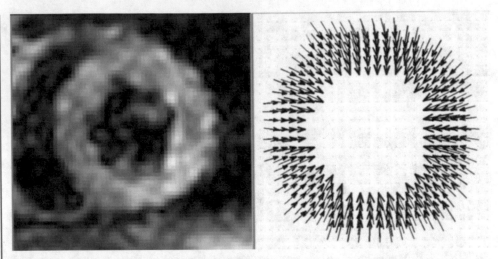

FIGURE 5.5: (Left) Mouse heart and corresponding (right) tagged image indicating motion field. The tagged image can be used in constructing the motion model for tracking the endocardial border.

constant cardinality (say C) by always choosing equal number of points from both sides of $[(X_i, Y_i)_{t-1}]$ on the trajectory. We now define the motion model as uniformly distributed over the trajectory:

$$p[(X_i, Y_i)_t | (X_i, Y_i)_{t-1}] = 1/C, \quad (X_i, Y_i)_t \in f[(X_i, Y_i)_{t-1}].$$

In order to execute MCMC in this context, we should also define the proposal density q from which samples for a snaxel are drawn. We know from the theory of the MCMC method that the most efficient sampling (fast convergence to stationary distribution) is obtained if the proposal density closely resembles the target stationary distribution [24]. We may define q as follows:

$$q[(X_i, Y_i)_t; (X_i, Y_i)_{t-1}] \propto e[(X_i, Y_i)_t], \quad (X_i, Y_i)_t \in f[(X_i, Y_i)_t],$$

where $e[(X_i, Y_i)_t]$ is the absolute value of the 1D directional edge strength at the point $[(X_i, Y_i)_t]$, the direction of the trajectory $f[(X_i, Y_i)_{t-1}]$ at $[(X_i, Y_i)_t]$. With these probability densities, the MH ratio now becomes

$$\text{MH} = \frac{p[Z_t|\{(X_j, Y_j)_t\}_{j=1, j\neq i}^{N}, (X, Y)]}{p[Z_t|\{(X_j, Y_j)_t\}_{j=1}^{N}]} \frac{\exp\left\{-\frac{d^2[(X,Y),(X_{i+1}, Y_{i+1})_t]}{2\sigma_s^2}\right\}}{\exp\left\{-\frac{d^2[(X_i, Y_i)_t,(X_{i+1}, Y_{i+1})_t]}{2\sigma_s^2}\right\}} \frac{e[(X_j, Y_j)_t]}{e[(X, Y)]},$$

where (X, Y) is the proposal point sampled from $q(\cdot)$ to be accepted against the current sample $(X_i, Y_i)_t$. Tracking examples using this approach are shown in Fig. 5.4.

5.4   SUMMARY

Shapes play a pivotal role in tracking biomedical objects. We have, in hand, an arsenal of shape-based approaches that accommodate both rigid and nonrigid targets. For rigid objects, we can apply the affine or projective snake trackers that adjust only a modicum of parameters to transform the contour in motion. When objects are not necessarily well modeled by rigid shapes, we can try the stochastic simulated annealing approach to tracking or a more expeditious deterministic annealing. Finally, the sequential Bayesian approach allows both shape-based object delineation, and incorporation of a stochastic motion model for tracking.

Biomedical image analysis is an expanding and active field. *Biomedical Image Analysis: Tracking* spotlights image analysis solutions where the goal is tracking a moving target. We began the book with an overview of active contour methods that have profoundly impacted the medical image analysis community and are now deployed in many working systems. Bayesian methods including Kalman filter-inspired solutions were described in the context of tracking in difficult biomedical imaging scenarios. Representing the latest frontier in biomedical image analysis, the particle filter and related methods were introduced and demonstrated. Finally, because we deal with remarkable shapes in the biomedical world, we described a group of shape-based tracking approaches. It is our hope that the reader can rapidly gain the ability to implement the tracking methods covered in this book for his/her own application.

Appendices

"Mathematics, rightly viewed, possess not only truth, but supreme beauty—a beauty cold and austere, like that of sculpture."

Betrand Russell

"Medicine is controls with poor actuators."

Peter Doerschuk

APPENDIX A: EULER AND GRADIENT DESCENT EQUATIONS FOR ACTIVE CONTOUR

To obtain the first variation of the snake energy functional (2.1) we follow the procedure of [2]. We consider a small neighborhood $[\varepsilon_1 U(s), \varepsilon_2 V(s)]$ of the contour $[X(s), Y(s)]$, where $U(s): \Re \to \Re$ and $V(s): \Re \to \Re$ are differentiable functions with $s \in [0, 1]$. After addition of the neighborhood to the contour, we obtain a perturbed contour, $[X(s) + \varepsilon_1 U(s), Y(s) + \varepsilon_2 V(s)]$. To maintain a closed contour after addition of the perturbation, we note that the following boundary conditions hold:

$$U(0) = U(1), \tag{A1}$$

and

$$V(0) = V(1). \tag{A2}$$

Additionally, we require that the perturbed contour be differentiable, so we also have

$$\frac{dU(0)}{ds} = \frac{dU(1)}{ds}, \tag{A3}$$

and

$$\frac{dV(0)}{ds} = \frac{dV(1)}{ds}. \tag{A4}$$

With the perturbation added to the contour the energy functional (2.1) is now a function of ε_1 and ε_2:

$$J(\varepsilon_1, \varepsilon_2) = \frac{1}{2} \int_0^1 \alpha \left[\left(\frac{dX}{ds} + \varepsilon_1 \frac{dU}{ds} \right)^2 + \left(\frac{dY}{ds} + \varepsilon_2 \frac{dV}{ds} \right)^2 \right] ds$$

$$+ \frac{1}{2} \int_0^1 \beta \left[\left(\frac{d^2X}{ds^2} + \varepsilon_1 \frac{d^2U}{ds^2} \right)^2 + \left(\frac{d^2Y}{ds^2} + \varepsilon_2 \frac{d^2V}{ds^2} \right)^2 \right] ds$$

$$- \int_0^1 f(X + \varepsilon_1 U, Y + \varepsilon_2 V) ds. \tag{A5}$$

In order to minimize (A5) where the minimizer is the unperturbed contour $[X(s), Y(s)]$, we need the following conditions [2]:

$$\left. \frac{\partial J}{\partial \varepsilon_1} \right|_{\substack{\varepsilon_1 = 0, \\ \varepsilon_2 = 0}} = 0, \tag{A6}$$

and

$$\left. \frac{\partial J}{\partial \varepsilon_2} \right|_{\substack{\varepsilon_1 = 0, \\ \varepsilon_2 = 0}} = 0. \tag{A7}$$

From (A5) we obtain

$$\left. \frac{\partial J}{\partial \varepsilon_1} \right|_{\substack{\varepsilon_1 = 0, \\ \varepsilon_2 = 0}} = \int_0^1 \left[\alpha \frac{dX}{ds} \frac{dU}{ds} + \beta \frac{d^2X}{ds^2} \frac{d^2U}{ds^2} - \frac{\partial f}{\partial x} U \right] ds. \tag{A8}$$

Applying the rule of integration by parts to the first and the second integrand of (A8) we have

$$\left. \frac{\partial J}{\partial \varepsilon_1} \right|_{\substack{\varepsilon_1 = 0, \\ \varepsilon_2 = 0}} = \alpha \frac{dX}{ds} [U(1) - U(0)] - \int_0^1 \alpha \frac{d^2X}{ds^2} U ds$$

$$+ \beta \left[\frac{dU(1)}{ds} - \frac{dU(0)}{ds} \right] - \int_0^1 \beta \frac{d^3X}{ds^3} \frac{dU}{ds} ds - \int_0^1 \frac{\partial f}{\partial x} U ds, \tag{A9}$$

where we have assumed C^3 continuity on the snake, *i.e.*, the snake itself, its first, second, and third derivative are all continuous; so we have $\frac{dX(0)}{ds} = \frac{dX(1)}{ds}$, $\frac{d^2X(0)}{ds^2} = \frac{d^2X(1)}{ds^2}$, and $\frac{d^3X(0)}{ds^3} = \frac{d^3X(1)}{ds^3}$ in particular. Now applying (A1) and (A3) to (A9) we obtain

$$\left.\frac{\partial J}{\partial \varepsilon_1}\right|_{\substack{\varepsilon_1=0,\\ \varepsilon_2=0}} = -\int_0^1 \alpha\frac{d^2X}{ds^2}U\,ds - \int_0^1 \beta\frac{d^3X}{ds^3}\frac{dU}{ds}\,ds - \int_0^1 \frac{\partial f}{\partial x}U\,ds, \qquad \text{(A10)}$$

If we once more apply the rule for integration by parts on (A10) we obtain

$$\left.\frac{\partial J}{\partial \varepsilon_1}\right|_{\substack{\varepsilon_1=0,\\ \varepsilon_2=0}} = -\int_0^1 \alpha\frac{d^2X}{ds^2}U\,ds - \beta\frac{d^3X}{ds^3}[U(1) - U(0)]$$

$$+ \int_0^1 \beta\frac{d^4X}{ds^4}U\,ds - \int_0^1 \frac{\partial f}{\partial x}U\,ds. \qquad \text{(A11)}$$

We reduce (A11) to the following by applying (A1):

$$\left.\frac{\partial J}{\partial \varepsilon_1}\right|_{\substack{\varepsilon_1=0,\\ \varepsilon_2=0}} = \int_0^1 \left[-\alpha\frac{d^2X}{ds^2} + \beta\frac{d^4X}{ds^4} - \frac{\partial f}{\partial x}\right]U\,ds. \qquad \text{(A12)}$$

Similarly utilizing (A2) and (A4) we obtain

$$\left.\frac{\partial J}{\partial \varepsilon_2}\right|_{\substack{\varepsilon_1=0,\\ \varepsilon_2=0}} = \int_0^1 \left[-\alpha\frac{d^2Y}{ds^2} + \beta\frac{d^4Y}{ds^4} - \frac{\partial f}{\partial y}\right]V\,ds. \qquad \text{(A13)}$$

From (A12) and (A13) the Euler equations (2.2) and (2.3) and the gradient descent equations (2.6) and (2.7) follow by their definition [2].

APPENDIX B: POSITIVE DEFINITE $(I_n + \zeta A)$

Let **x** denote the following vector:

$$\mathbf{x} = [X_0, X_1, \ldots, X_{n-1}]^T,$$

then, we can write

$$\mathbf{x}^T A \mathbf{x} = \sum_{i=0}^{n-1} -\alpha(X_{i+1} - 2X_i + X_{i-1})X_i + \beta(X_{i+2} - 4X_{i+1}$$
$$+6X_i - 4X_{i-1} + X_{i-2})X_i. \tag{B1}$$

We rewrite (B1) in the following way:

$$\mathbf{x}^T A \mathbf{x} = \alpha \sum_{i=0}^{n-1} \left(X_i^2 - 2X_i X_{i+1} + X_{i+1}^2 \right) + \beta \sum_{i=0}^{n-1} \left(X_i^2 - 2X_i X_{i+2} + X_{i+2}^2 \right)$$
$$+4\beta \sum_{i=0}^{n-1} \left(X_i^2 - 2X_i X_{i+1} + X_{i+1}^2 \right)$$
$$= \alpha \sum_{i=0}^{n-1} (X_i - X_{i+1})^2 + \beta \sum_{i=0}^{n-1} (X_i - X_{i+2})^2$$
$$+4\beta \sum_{i=0}^{n-1} (X_i - X_{i+1})^2 \geq 0, \quad \forall \mathbf{x}. \tag{B2}$$

Therefore from (B2) we conclude that the matrix A is non-negative definite. Since ζ is a positive number, the matrix ζA is also non-negative definite. Thus adding I_n to it makes the sum positive definite, because I_n itself is positive definite.

APPENDIX C: DERIVATION OF EULER EQUATIONS FOR GRADIENT VECTOR FLOW

The first variations of (2.33) with respect to u and v are obtained by adding small perturbations functions $\alpha p(x, y)$ and $\beta q(x, y)$ respectively to $u(x, y)$ and $v(x, y)$, then taking derivatives with respect to α and β. So, for α we obtain

$$\lim_{\alpha \to 0} \frac{E_{\text{GVF}}(u + \alpha p, v) - E_{\text{GVF}}(u, v)}{\alpha} = \iint [\mu(u_x p_x + u_y p_y)$$
$$+(f_x^2 + f_y^2)(u - f_x)p]dxdy. \tag{C1}$$

Applying Green's theorem to (C1) we obtain

$$\lim_{\alpha \to 0} \frac{E_{\text{GVF}}(u + \alpha p, v) - E_{\text{GVF}}(u, v)}{\alpha} = \mu \left[\int_{\partial\Omega} p(\nabla u \cdot d\sigma) - \iint_{\Omega} p \nabla^2 u \, dxdy \right]$$
$$+ \iint_{\Omega} (f_x + f_y)(u - f_x)p \, dxdy, \tag{C2}$$

where Ω is the 2D image domain and $\partial\Omega$ is the boundary of Ω, and $d\sigma$ a small element (in this case line segment) of $\partial\Omega$. Applying the "natural" boundary condition that ∇u is zero at boundary $\partial\Omega$, the first integration vanishes and we are left with

$$\lim_{\alpha \to 0} \frac{E_{\mathrm{GVF}}(u + \alpha p, v) - E_{\mathrm{GVF}}(u, v)}{\alpha}$$
$$= -\iint_{\Omega} \left[\mu\nabla^2 u - (f_x + f_y)(u - f_x)\right] p \, dx dy. \tag{C3}$$

From (C3) the first variation of the GVF is as follows:

$$\frac{\delta E_{\mathrm{GVF}}}{\delta u} = -[\mu\nabla^2 u - (f_x + f_y)(u - f_x)]. \tag{C4}$$

From (C4) the Euler equations (2.34) and (2.35) follow.

APPENDIX D: GRADIENT DESCENT EQUATIONS FOR RADIAL SNAKE

In the following we denote $R(t)$ by R for simplicity. The first variation [2] of E_{edge} in (2.39), with respect to P, is obtained simply by differentiating (2.39) with respect to P, as P is a variable here, and not a function:

$$\frac{\delta E_{\mathrm{r-snake}}}{\delta P} = -\frac{\partial}{\partial P} \left\{ \frac{\int_0^{2\pi} w\left[P + R\cos(t),\, Q + R\sin(t)\right] R \, dt}{L_s} \right\}$$

$$= -\frac{\int_0^{2\pi} \frac{\partial}{\partial x} w\left[P + R\cos(t),\, Q + R\sin(t)\right] R \, dt}{L_s}$$

$$= -\frac{\int_0^{2\pi} \left\{ \frac{\partial w}{\partial x}\left[P + R\cos(t),\, Q + R\sin(t)\right] \right\} R \, dt}{L_s}$$

$$= -\overline{w_x}.$$

In a similar fashion, we obtain

$$\frac{\delta E_{\mathrm{r-snake}}}{\delta Q} = -\overline{w_y} + \mu_{\mathrm{pos}}(Q - P_y).$$

The first variation of E_{edge} with respect to R is as follows:

$$\frac{\delta E_{\text{r-snake}}}{\delta R} = -\frac{1}{L_s^2}\left(L_s\frac{\delta E}{\delta R} - E\frac{\delta L_s}{\delta R}\right) + \mu_{\text{cons}}(R - \rho)$$

$$= -\frac{1}{L_s}\left(\frac{\delta E}{\delta R} - \overline{w}\frac{\delta L_s}{\delta R}\right) + \mu_{\text{cons}}(R - \rho),$$

where L_s is given as

$$L_s = \int\limits_0^{2\pi} R(t)dt,$$

and, $\overline{w}$ is given in (2.45) and, E is as follows:

$$E = \int\limits_0^{2\pi} w[P + R\cos(t),\, Q + R\sin(t)]\, R dt.$$

So to obtain the first variation of E with respect to R, we add a small perturbation $\varepsilon S(t)$ to $R(t)$, and take the derivative with respect to ε:

$$\lim_{\varepsilon \to 0} \frac{E(P, Q, R + \varepsilon S) - E(P, Q, R)}{\varepsilon}$$

$$= \lim_{\varepsilon \to 0} \frac{1}{\varepsilon}\left\{ \begin{array}{l} \int\limits_0^{2\pi} w[P + R\cos(t) + \varepsilon S\cos(t),\, Q + R\sin(t) + \varepsilon S\sin(t)](R + \varepsilon S)dt \\[2mm] - \int\limits_0^{2\pi} w[P + R\cos(t),\, Q + R\sin(t)]\, R dt \end{array} \right\}$$

$$= \lim_{\varepsilon \to 0} \frac{1}{\varepsilon}\left(\int\limits_0^{2\pi} \{w[P + R\cos(t) + \varepsilon S\cos(t),\, Q + R\sin(t) + \varepsilon S\sin(t)]\right.$$

$$\left. - w[P + R\cos(t),\, Q + R\sin(t)]\}\, R dt\right)$$

$$+ \lim_{\varepsilon \to 0} \int\limits_0^{2\pi} w[P + R\cos(t) + \varepsilon S\cos(t),\, Q + R\sin(t) + \varepsilon S\sin(t)]\, S dt$$

$$= \int\limits_0^{2\pi}\left\{ S\cos(t)\frac{\partial}{\partial x}w[P + R\cos(t),\, Q + R\sin(t)] \right.$$

$$\left. + S\sin(t)\frac{\partial}{\partial y}w[P + R\cos(t),\, Q + R\sin(t)] \right\} R dt$$

$$+ \int_0^{2\pi} w[P + R\cos(t),\ Q + R\sin(t)]S dt$$

$$= \int_0^{2\pi} \left\{ \left[\frac{\partial w}{\partial x} \cos(t) + \frac{\partial w}{\partial y} \sin(t) \right] R + w \right\} S dt.$$

So we now have

$$\frac{\delta E}{\delta R} = w + R\frac{\partial w}{\partial x} \cos(t) + R\frac{\partial w}{\partial y} \sin(t).$$

The first variation of L_s is obtained in a similar way

$$\lim_{\varepsilon \to 0} \frac{1}{\varepsilon} \left(\int_0^{2\pi} ((R + \varepsilon S) - R) dt \right) = \int_0^{2\pi} S dt.$$

So the first variation of L_s is given by $\frac{\delta L_s}{\delta R} = 1$, and thus, we finally obtain

$$\frac{\delta E_{\text{r-snake}}}{\delta R} = -\frac{1}{L_s} \left[w + R\frac{\partial w}{\partial x} \cos(t) + R\frac{\partial w}{\partial y} \sin(t) - \overline{w} \right] + \mu_{\text{cons}}(R - \rho).$$

Finally, we obtain the gradient descent equation (2.38) for P as follows:

$$\frac{\partial P}{\partial \tau} = -\frac{\delta E_{\text{r-snake}}}{\delta P} = -\frac{\delta E_{\text{edge}}}{\delta P} - \mu_{\text{cons}} \frac{\delta E_{\text{cons}}}{\delta P} - \mu_{\text{pos}} \frac{\delta E_{\text{pos}}}{\delta P} = \overline{w_x}.$$

In a similar manner we derive (2.43) and (2.44).

APPENDIX E: MINIMIZATION OF MGVF ENERGY FUNCTIONAL

Minimization of the MGVF energy functional can be achieved with variational calculus as follows. The first variation of (2.48) with respect to w is obtained by adding a small perturbation function $\alpha q(x, y)$ to $w(x, y)$, and taking the derivative with respect to α:

$$
\lim_{\alpha \to 0} \frac{E_{\mathrm{MGVF}}(w + \alpha q) - E(w)}{\alpha}
$$

$$
= \lim_{\alpha \to 0} \frac{1}{2\alpha} \left[\begin{pmatrix} \iint \mu |\nabla w|^2 \{ H_\varepsilon[\nabla w \cdot (v^x, v^y) + \alpha \nabla q \cdot (v^x, v^y)] \\ - H_\varepsilon[\nabla w \cdot (v^x, v^y)]\} dxdy \\ + \alpha \iint \{\mu H_\varepsilon[\nabla w \cdot (v^x, v^y) + \alpha \nabla q \cdot (v^x, v^y)]\nabla w . \nabla q\} dxdy \\ + \frac{\alpha^2}{2} \iint \{\mu H_\varepsilon[\nabla w \cdot (v^x, v^y) + \alpha \nabla q \cdot (v^x, v^y)]|\nabla q|^2\} dxdy \\ + \alpha \iint [f(w - f)q]dxdy + \frac{\alpha^2}{2} \iint (fq^2)dxdy \end{pmatrix} \right].
$$

$$\tag{E1}$$

Now, applying the Mean Value Theorem [41]:

$$
H_\varepsilon[\nabla w \cdot (v^x, v^y) + \alpha \nabla q \cdot (v^x, v^y)] - H_\varepsilon[\nabla w \cdot (v^x, v^y)]
$$
$$
= \alpha \nabla q \cdot (v^x, v^y) H'_\varepsilon[\nabla w \cdot (v^x, v^y) + \alpha \theta \nabla q \cdot (v^x, v^y)],
$$

$$\tag{E2}$$

where $\theta(x, y)$ is a function with $0 < \theta(x, y) < 1, \forall x, y$. Using (E2) in (E1), we obtain

$$
\lim_{\alpha \to 0} \frac{E_{\mathrm{MGVF}}(w + \alpha q) - E(w)}{\alpha} = \frac{1}{2} \iint \mu \{|\nabla w|^2 H'_\varepsilon[\nabla w \cdot (v^x, v^y)]\nabla q
$$
$$
\cdot (v^x, v^y)\} dxdy + \iint \{\mu H_\varepsilon[\nabla w \cdot (v^x, v^y)]\nabla w \cdot \nabla q\} dxdy + \iint [f(w - f)q]dxdy,
$$

$$\tag{E3}$$

where

$$
H'_\varepsilon(z) = \frac{dH(z)}{dz} = \frac{1}{\pi} \left(\frac{\varepsilon}{z^2 + \varepsilon^2} \right).
$$

$$\tag{E4}$$

Now the first integral in (E3) can be written as

$$
\left| \iint \mu \{|\nabla w|^2 H'_\varepsilon[\nabla w . (v^x, v^y)]\nabla q \cdot (v^x, v^y)\} dxdy \right|
$$
$$
= \left| \frac{\mu}{\pi} \iint \frac{\varepsilon |\nabla w|^2 \nabla q \cdot (v^x, v^y)}{|\nabla w . (v^x, v^y)|^2 + \varepsilon^2} dxdy \right|
$$
$$
\leq \frac{\varepsilon \mu}{\pi} \iint \frac{|\nabla w|^2 |\nabla q \cdot (v^x, v^y)|}{|\nabla w . (v^x, v^y)|^2} dxdy.
$$

$$\tag{E5}$$

Except when the gradient of w is perpendicular or nearly perpendicular to the target object velocity (v^x, v^y), the contribution of the first integral in (E3) is negligible for small values of ε. On the other hand, the minimization of MGVF energy functional (2.48) should lead to small $|\nabla w|$, when ∇w is perpendicular or nearly perpendicular to the object velocity. In other words, we may assume that when $|\nabla w.(v^x, v^y)|$ is infinitesimal so is $|\nabla w|$, $i.e.$ $\frac{|\nabla w|}{|\nabla w.(v^x, v^y)|} \approx 1$ for small $|\nabla w.(v^x, v^y)|$. In this case we have

$$\frac{\varepsilon\mu}{\pi} \iint \frac{|\nabla w|^2 |\nabla q \cdot (v^x, v^y)|}{|\nabla w.(v^x, v^y)|^2} dxdy \approx \frac{\varepsilon\mu}{\pi} \iint |\nabla q \cdot (v^x, v^y)| dxdy. \tag{E6}$$

From (E5) and (E6) we observe that for small values of ε the first integral in (E3) is negligible when ∇w is perpendicular or nearly perpendicular to the flow velocity. Therefore we neglect the contribution of the first integral of (E3) whether or not ∇w is perpendicular to the flow velocity and now we are only left with the second and the third integral of (E3).

Applying Green's theorem [41] to the second integral of (E3) we obtain

$$\iint \{\mu H_\varepsilon[\nabla w.(v^x, v^y)]\nabla w.\nabla q\} dxdy = \int_{\partial\Omega} \mu q H_\varepsilon[\nabla w.(v^x, v^y)]\nabla w \cdot d\sigma$$
$$- \iint \mu q \, div\{H_\varepsilon[\nabla w.(v^x, v^y)]\nabla w\} dxdy. \tag{E7}$$

Thus after disregarding the first integral of (E3), combining it with (E7) and applying the boundary condition $\nabla w \cdot \mathbf{n}_{\partial\Omega} = 0$ on $\partial\Omega$, where $\mathbf{n}_{\partial\Omega}$ denotes the perpendicular to the boundary $\partial\Omega$, we obtain

$$\lim_{\alpha\to 0} \frac{E_{\mathrm{MGVF}}(w + \alpha q) - E(w)}{\alpha} = \iint (f(w - f)$$
$$-\mu div\{H_\varepsilon[\nabla w.(v^x, v^y)]\nabla w\})q \, dxdy. \tag{E8}$$

So the first variation of the energy functional (2.48) obtained from (E8) is follows:

$$\frac{\delta E_{\mathrm{MGVF}}}{\delta w} = -\mu \, div\{H_\varepsilon[\nabla w.(v^x, v^y)]\nabla w\} + f(w - f). \tag{E9}$$

And now the gradient descent equation to obtain the minimizing function w is as follows:

$$\frac{\partial w}{\partial \tau} = -\frac{\delta E_{\text{MGVF}}}{\delta w} = \mu \, \text{div}\{H_\varepsilon[\nabla w.(v^x, v^y)]\nabla w\} - f(w - f).$$

APPENDIX F: CONVERGENCE OF NUMERICAL IMPLEMENTATION OF MGVF

If r denotes the location (i, j), then (2.51) can be rewritten as

$$w_r^{\tau+1} = \left(1 - \frac{f_r}{\lambda} - \frac{\mu}{\lambda}\sum_{q \in N(r)} c_q^\tau\right) w_r^\tau + \frac{\mu}{\lambda}\sum_{q \in N(r)} c_q^\tau w_q^\tau + \frac{f_r^2}{\lambda}, \qquad \text{(F1)}$$

where $N(r)$ denotes the eight neighbors of the location r and c_q^τ is as follows:

$$c_q^\tau = H_\varepsilon\left[(lv^x + mv^y)(w_q^\tau - w_r^\tau)\right], \qquad \text{(F2)}$$

where (l, m) denotes the vector representing the direction from location r to the location q in $N(r)$, and (v^x, v^y) denotes the leukocyte velocity direction. Let us now impose the following constraint:

$$\lambda \geq \max_r \left(f_r + \mu \sum_{q \in N(r)} c_q^\tau\right). \qquad \text{(F3)}$$

Then, all the coefficients of ws in (F1) become non-negative. Equation (F1) can be rewritten for all the grid locations of the image domain in a matrix-vector form as follows:

$$\mathbf{w}^{\tau+1} = A^\tau \mathbf{w}^\tau + \mathbf{f}, \qquad \text{(F4)}$$

where A^τ is a $M \times M$ matrix, M being the total number of grid points in the image domain. A^τ has all non-negative elements and any of its row contains at the most nine positive elements. Vectors $\mathbf{f}$ and $\mathbf{w}^\tau$ are column vectors given as follows:

$$\mathbf{f} = [\frac{f_1^2}{\lambda}\cdots\frac{f_M^2}{\lambda}]^T, \quad \mathbf{w}^\tau = [w_1^\tau\cdots w_M^\tau]^T. \qquad \text{(F5)}$$

We can rewrite (F4) as follows:

$$\mathbf{w}^{\tau+1} = A^{\tau} A^{\tau-1} \cdots A^0 \mathbf{w}^0 + (A^{\tau} A^{\tau-1} \cdots A^1)\mathbf{f}$$
$$+ (A^{\tau} A^{\tau-1} \cdots A^2)\mathbf{f} + \ldots + (A^{\tau})\mathbf{f} + \mathbf{f}. \qquad (F6)$$

If we now impose condition (2.52) on the edge-map, then it can be shown that,

$$(\mathbf{f})_r \leq \frac{1}{\lambda}, \ (A^{\tau}\mathbf{f})_r \leq (1 - \frac{\delta}{\lambda})\frac{1}{\lambda}, \cdots\cdots, \left[(A^{\tau} A^{\tau-1} \cdots A^{\tau-i+1})\mathbf{f}\right]_r$$
$$\leq (1 - \frac{\delta}{\lambda})^i \frac{1}{\lambda}, \quad \forall r,$$

where $(\mathbf{x})_r$ denotes the rth element of the vector $\mathbf{x}$. Now, note that (F6) is a series with non-negative terms, so by the comparison test with a geometric series of common ratio: $0 < (1 - (\delta/\lambda)) < 1$, (F6) converges as $\tau \to \infty$. Finally from (F2) we note that the maximum value a coefficient, c_q^{τ}, can achieve is unity; also the maximum value of the normalized edge-map f_r is unity. The right-hand side of (F3) never exceeds $1 + 8\mu$ assuming an eight-neighborhood; consequently we set λ to $1 + 8\mu$ in the numerical implementation of (2.51).

References

[1] M. Kass, A. Witkin, and D. Terzopoulos, "Snakes: Active contour models," *Int. J. Computer Vision*, pp. 321–331, 1987.

[2] R. Courant and D. Hilbert, *Methods of Mathematical Physics*, vol. 1. New York: Interscience, 1953.

[3] J. L. Troutman, *Variational Calculus with Elementary Convexity*, New York: Springer-Verlag, 1983.

[4] C. A. Hall and T. A. Porsching, *Numerical Analysis of Partial Differential Equations*. Englewood Cliffs, NJ: Prentice Hall, 1990.

[5] N. Ray, B. Chanda, and J. Das, "A fast and flexible multiresolution snake with a definite termination criterion," *Pattern Recognition*, vol. 34, pp. 1483–1490, 2001. doi:10.1016/S0031-3203(00)00077-7

[6] L. D. Cohen, "On active contour models and balloons," *CVGIP: Image Understanding*, vol. 53, pp. 211–218, 1991. doi:10.1016/1049-9660(91)90028-N

[7] N. Ray, S. T. Acton, T. Altes, E. E. de Lange, and J. R. Brookeman, "Merging parametric active contours within homogeneous image regions for MRI-based lung segmentation," *IEEE Trans. Med. Imag.*, vol. 22, pp. 189–199, 2003. doi:10.1109/TMI.2002.808354

[8] C. Xu and J. L. Prince, "Snakes, shapes, and gradient vector flow," *IEEE Trans. Image Process.*, vol. 7, pp. 359–369, 1998. doi:10.1109/83.661186

[9] N. Ray, S. T. Acton, "Motion gradient vector flow: an external force for tracking rolling leukocytes with shape and size constrained active contours," *IEEE Trans. Med. Imaging*, vol. 23, no.12, pp. 1466–1478, 2004. doi:10.1109/TMI.2004.835603

[10] N. Ray, S. T. Acton, and K. Ley, "Tracking leukocytes in vivo with shape and size constrained active contours," *IEEE Trans. Med. Imaging*, vol. 21, pp. 1222–1235, 2002. doi:10.1109/TMI.2002.806291

[11] N. Nordstrom, "Biased anisotropic diffusion: A unified regularization and diffusion approach to edge detection," *Image Vis. Comput.*, vol. 8, pp. 318–327, 1990. doi:10.1016/0262-8856(90)80008-H

[12] M. A. Gennert and A. Y. Yuille, "Determining the optimal weights in multiple objective function optimization," in *Proc. 2nd Int. Conf. Computer Vision*, pp. 87–89, 1988.

[13] W. K. Pratt, *Digital Image Processing*. New York: John Willy, 1981.

[14] A. A. Amini, T. E. Weymouth, and R. C. Jain, "Using dynamic programming for solving variational problems in vision," *IEEE Trans. Patt. Anal. Mach. Intell.*, vol. 12, no. 9, pp. 855–867, 1990. doi:10.1109/34.57681

[15] M. S. Arulampalam, S. Maskell, N. Gordon, and T. Clapp, "A tutorial on particle filters for online nonlinear/non-Gaussian Bayesian tracking," *IEEE Trans. Signal Proc.*, vol. 50, pp. 174–188, 2002. doi:10.1109/78.978374

[16] H. V. Poor, *An Introduction to Signal Detection and Estimation*. New York: Springer, 1998.

[17] S. Blackman and R. Popoli, *Design and Analysis of Modern Tracking Systems*, Boston, MA: Artech House, 1999.

[18] H. A. P. Blom and Y. Bar-Shalom, "The interacting multiple model algorithm for systems with Markovian switching coefficients," *IEEE Trans. Autom. Control*, vol. 33, no. 8, pp. 780–783, 1988. doi:10.1109/9.1299

[19] A. Blake and M. Isard, *Active Contours: The Application of Techniques from Graphics, Vision, Control Theory and Statistics to Visual Tracking of Shapes in Motion*. Berlin: Springer-Verlag, 1998.

[20] U. Grenander, Y. Chow, and D. M. Keenan, *HANDS: A Pattern Theoretical Study of Biological Shapes*. New York: Springer-Verlag, 1991.

[21] A. Doucet, N. de Freitas, and N. Gordon, *Sequential Monte Carlo Methods in Practice*. New York: Springer-Verlag, 2001.

[22] L. D. Stone, C. A. Barlow, and T. L. Corwin, *Bayesian Multiple Target Tracking*. Boston, MA: Artech House, 1999.

[23] Y. Bar-Shalom, X. R. Li, *Multitarget-Multisensor Tracking: Principles and Techniques*. Storrs, CT: YBS Publishing, 1995.

[24] J. S. Liu, *Monte Carlo Strategies in Scientific Computing*. New York: Springer, 2001.

[25] D. Bertsekas, *Network Optimization: Continuous and Discrete Methods*. Athena Scientific, Nashua, NH, 1998.

[26] D. J. Best and N. I. Fisher, "Efficient simulation of the von Mises distribution," *Appl. Stat.*, vol. 28, no. 2, pp. 152–157, 1979.

[27] G. Dong, N. Ray, and S. T. Acton, "Intravital leukocyte detection using the gradient inverse coefficient of variation," *IEEE Trans. Medical Imag.*, vol. 24, pp. 910–924, 2005. doi:10.1109/TMI.2005.846856

[28] M. Abramowitz and I. Stegun, *Handbook of Mathematical Functions*. Washington, D.C.: National Bureau of Standards, 1966.

[29] M. E. Welser and S. T. Acton, "Projection model snakes for tracking using a Monte Carlo approach," *J. Electron. Imag.*, vol. 13, pp. 384–398, 2004. doi:10.1117/1.1687732

[30] J. Besag, "Markov chain Monte Carlo for statistical inference," University of Washington, Seattle, USA, September 2000.

[31] S. P. Brooks, P. Dellaportas, and G. O. Roberts, "An approach to diagnosing total variation convergence of MCMC algorithms," *J. Comput. Graphical Stat.*, vol. 6, pp. 251–265, 1997.

[32] A. R. Mansouri, D. P. Mukerjee, and S. T. Acton, "Constraining active contour evolution via Lie groups of transformation," *IEEE Trans. Image Process.*, vol. 13, no. 6, pp. 853–863, 2004. doi:10.1109/TIP.2004.826128

[33] N. S. Friedland and A. Rosenfeld, "Compact object recognition using energy-function-based optimization," *IEEE Trans. Pattern Anal. Mach. Intell.*, vol. 14, pp. 770–777, 1992. doi:10.1109/34.142912

[34] R. P. Grzeszczuk and D. N. Levin, "Browian strings: segmenting images with stochastically deformable contours," *IEEE Trans. Pattern Anal. Mach. Intell.*, vol. 19, pp. 1100–1114, 1997. doi:10.1109/34.625111

[35] G. Storvik, "A Bayesian approach to dynamic contours through stochastic sampling and simulated annealing," *IEEE Trans. Pattern Anal. Mach. Intell.*, vol. 16, pp. 976–986, 1994. doi:10.1109/34.329011

[36] D. Geman and S. Geman, "Stochastic relaxation, Gibbs distributions, and Bayesian restoration of images," *IEEE Trans. Pattern Anal. Machine Intell.*, vol. PAMI-6, pp. 721–741, 1984.

[37] E. H. L. Aarts and J. Korst, *Simulated Annealing and Boltzmann Machines: A Stochastic Approach to Combinatorial Optimization and Neural Computing.* New York: Wiley, 1987.

[38] S. T. Acton and A. C. Bovik, "Generalized deterministic annealing," *IEEE Trans. Neural Networks*, vol. 7, pp. 686–699, 1996. doi:10.1109/72.501726

[39] S. T. Acton, "Snakes for tracking via generalized deterministic annealing," *J. Electron. Imag.*, vol. 14, pp. 1–13, 2005.

[40] S. Z. Li, *Markov Random Field Modeling in Image Analysis.* New York: Springer, 2001.

[41] W. Rudin, *Principles of Mathematical Analysis*, 3rd ed. New York: McGraw-Hill, 1976.

Biographies

Scott T. Acton received the M.S. degree in electrical and computer engineering and the Ph.D. degree in electrical and computer engineering from the University of Texas at Austin in 1990 and 1993, respectively, where he was a Microelectronics and Computer Development Fellow. He received the B.S. degree in electrical engineering from Virginia Tech, Blacksburg in 1988 as a Virginia Scholar and Marshall Hahn Fellow. He was the class of 1984 valedictorian at Oakton High School in Vienna, VA. At Oakton, he was a member of the football and basketball teams, although the coach rarely let him in an actual game.

He has worked in industry for AT&T, Oakton, VA, the MITRE Corporation, McLean, VA, and Motorola, Inc., Phoenix, AZ and in academia for Oklahoma State University, Stillwater. Currently, he holds the Walter N. Munster Chair for Intelligence Enhancement at the University of Virginia (U.Va.), where he is a member of the Charles L. Brown Department of Electrical and Computer Engineering and the Department of Biomedical Engineering. At U.Va., he was named the Outstanding New Teacher in 2002 and was elected a Faculty Fellow in 2003. For his research in video tracking, he was given an ARO Young Investigator Award. He received the Halliburton Outstanding Young Faculty Award in 1998. In 1997, he was named the Eta Kappa Nu Outstanding Young Electrical Engineer—a national award that has been given annually since 1936. He is the recipient of a Whitaker Foundation Biomedical Engineering Research Grant for work in cell detection and tracking. He currently is working on NIH-funded research for leukocyte tracking, segmentation of the myocardium from ultrasound, and segmentation of the myocardium from MRI.

Dr. Acton is an active participant in the IEEE, served as Associate Editor for the *IEEE Transactions on Image Processing* and as Associate Editor for the *IEEE Signal Processing Letters*. He is the 2004 Technical Program Chair and the 2006 General Chair for the *Asilomar Conference on Signals, Systems and Computers*. His research interests include anisotropic diffusion, the history of Billy the Kid, active contours, biomedical segmentation problems, biomedical tracking problems, and war.

Acton was a 2005 finalist in the Florida First Coast Novel Contest. He lives in beautiful Charlottesville, Virginia, with his wife and two boys.

Nilanjan Ray received Bachelor of Mechanical Engineering from Jadavpur University, Calcutta, India in 1995; Master of Technology in computer science from Indian Statistical Institute, Calcutta, India in 1997; and Doctor of Philosophy in electrical engineering from the University of Virginia, Charlottesville, in 2003.

From 2003 to 2005, he held a postdoctoral researcher position at C.L. Brown Department of Electrical and Computer Engineering, University of Virginia. Currently he is a senior research scientist at UtopiaCompression Corporation, Los Angeles, CA.

Dr. Ray is a recipient of CIMPA-UNESCO fellowship for image processing school in 1999, graduate fellowship at Indian Statistical Institute from 1995 to 1997, best student paper award at IEEE international conference on image processing, held at Rochester, NY, 2002, from the IBM Picture Processing Society. Dr. Ray's research interests include object tracking using sequential Monte Carlo methods, applications of partial differential equations in image and video processing, image registration, and image denoising.

Dr. Ray lives on the beach in LA with his wife and new baby boy.